Copyright © 2024 by Charlie O. Williams

All rights reserved. No part of this publication may be reproduced, distributed, or transmitted in any form or by any means, including photocopying, recording, or other electronic or mechanical methods, without the prior written permission of the publisher, except in the case of brief quotations embodied in critical reviews and certain other noncommercial uses permitted by copyright law.

For permission requests, write to the publisher, addressed "Attention: Permissions Coordinator," at the address below.

[335 Glyn St, Glenstantia, Gauteng, South Africa]

Preface:

Welcome to "The Universe and Cosmology: Explore the Mysteries of the Cosmos, from the Big Bang Theory to the Latest Discoveries in Astronomy and Astrophysics." In this book, we embark on an exhilarating journey through the vast expanse of the universe, delving into the fascinating realms of cosmology, astronomy, and astrophysics.

The cosmos has always captivated the human imagination, inspiring awe and wonder with its beauty, complexity, and mystery. From the ancient civilizations that gazed up at the stars in wonder to the modern scientists who probe the depths of space with cutting-edge technology, humanity's quest to understand the universe has been a journey of discovery, exploration, and revelation.

In this book, we aim to share the excitement and wonder of cosmological exploration with readers of all backgrounds and interests. Whether you are a seasoned enthusiast or a curious newcomer to the field, there is something here for everyone to discover and enjoy.

We begin our journey with an exploration of the Big Bang theory—the prevailing scientific model for the origin and evolution of the universe. From the fiery birth of the cosmos to the formation of galaxies, stars, and planets, we trace the epic story of cosmic evolution and uncover the secrets of the universe's earliest moments.

As we journey deeper into the cosmos, we encounter a rich tapestry of phenomena, theories, and discoveries that challenge our understanding of the universe and push the boundaries of scientific knowledge. From the enigmatic nature of dark matter and dark energy to the cataclysmic collisions of black holes and the cosmic symphony of gravitational waves, we explore the most profound mysteries of the cosmos and the cutting-edge research that seeks to unravel them.

But our journey is not just about the science—it is also about the human spirit of exploration, curiosity, and wonder. As we gaze up at the night sky, we are reminded of our place in the cosmos and the interconnectedness of all things. The universe is not just a collection of stars and galaxies; it is a reflection of the human quest for understanding and discovery.

We hope that this book will inspire readers to embark on their own cosmic odyssey, to look up at the stars with wonder and curiosity, and to explore the mysteries of the universe with an open mind and a sense of awe. Whether you are a scientist, a student, or simply a lover of the cosmos, there is always something new to discover and learn about the universe we call home.

So join us as we journey through the cosmos, from the Big Bang to the latest discoveries in astronomy and astrophysics. The universe awaits, ready to reveal its secrets to those who dare to explore its depths.

TABLE OF CONTENT

Preface:
Introduction:
Chapter 1: The Big Bang Theory: Origin of the Universe
Chapter 2: Early Universe: Primordial Soup
Chapter 3: Formation of Galaxies and Stars
Chapter 4: Stellar Evolution: Birth, Life, and Death of Stars
Chapter 5: Black Holes: Nature's Cosmic Vacuum Cleaners
Chapter 6: Exoplanets: Worlds Beyond Our Solar System
Chapter 7: Dark Matter: Unseen Influences in the Cosmos
Chapter 8: Dark Energy: Mysterious Forces Shaping the Universe
Chapter 9: The Structure of the Universe: Clusters, Superclusters, and Filaments
Chapter 10: Cosmic Microwave Background Radiation: Echoes of the Big Bang
Chapter 11: Galaxy Formation and Evolution

Chapter 12: The Expanding Universe: Hubble's Law and Beyond
Chapter 13: Multiverse Theory: Exploring Other Realms
Chapter 14: Gravitational Waves: Ripples in Spacetime
Chapter 15: Time and Space: Einstein's Theory of Relativity
Chapter 16: Cosmic Inflation: Rapid Expansion in the Early Universe
Chapter 17: Quantum Cosmology: Bridging the Gap Between Micro and Macro Worlds
Chapter 18: Future Directions in Cosmology: Challenges and Discoveries Ahead
Conclusion:
Acknowledgments:

The Universe and Cosmology:
Explore the mysteries of the cosmos,
from the Big Bang theory to the latest
discoveries in astronomy and
astrophysics

Introduction:

Welcome to "The Universe and Cosmology: Explore the mysteries of the cosmos, from the Big Bang theory to the latest discoveries in astronomy and astrophysics." In this comprehensive exploration, we embark on a journey through the vast expanse of the universe, delving into the origins, structure, and evolution of the cosmos. From the enigmatic beginnings of the Big Bang to the cutting-edge research reshaping our understanding of space and time, join us as we unravel the secrets of the universe and marvel at the wonders that lie beyond the reaches of our imagination. Whether you're a seasoned astronomer or a curious observer, prepare to be captivated by the awe-inspiring beauty and complexity of the cosmos.

Chapter 1: The Big Bang Theory: Origin of the Universe

In the annals of cosmology, few concepts rival the sheer magnitude and significance of the Big Bang theory. It stands as the cornerstone of modern cosmological understanding, providing a framework for comprehending the origin, evolution, and structure of the universe itself. From the primordial singularity to the cosmic microwave background radiation, the journey of the cosmos unfolds through the lens of this paradigm-shifting concept.

1.1 The Genesis of the Big Bang Theory

The seeds of the Big Bang theory were sown in the early 20th century, amidst a fervent period of scientific inquiry and discovery. Albert Einstein's revolutionary theory of general relativity, published in 1915, laid the groundwork for understanding the nature of gravity and its effects on the fabric of spacetime. However, it would be another decade before its implications for the universe as a whole were fully realized.

1.1.1 Einstein's Static Universe

Initially, Einstein himself envisioned a static, unchanging cosmos, a universe that existed eternally without beginning or end. To maintain such a state, he introduced the cosmological constant into his equations—a term intended to counteract the gravitational attraction between matter and prevent the universe from collapsing under its own weight.

1.1.2 The Work of Georges Lemaître

It was the Belgian physicist and Catholic priest Georges Lemaître who first proposed the idea of an expanding universe in 1927. Building upon Einstein's equations, Lemaître's model described a dynamic cosmos, one in which galaxies were receding from each other as spacetime itself expanded. Despite initial skepticism, Lemaître's insights would prove pivotal in the development of the Big Bang theory.

1.2 Hubble's Cosmic Revelation

The turning point came in 1929, with the groundbreaking observations of American astronomer Edwin Hubble. Through meticulous observations of distant galaxies using the newly constructed Hooker Telescope at Mount Wilson Observatory, Hubble made a startling discovery: galaxies beyond our own Milky Way were not only moving away from us but were receding at velocities proportional to their distance.

1.2.1 Hubble's Law and the Expanding Universe

Hubble's observations revealed a direct relationship between the distance to a galaxy and its recessional velocity, a relationship encapsulated in what would come to be known as Hubble's Law. This empirical evidence provided compelling support for the idea of cosmic expansion and dealt a decisive blow to the notion of a static universe.

1.3 The Birth of the Big Bang Theory

Building upon the work of Lemaître and Hubble, a coherent picture of cosmic evolution began to emerge. In 1931, the term "Big Bang" was coined by British astrophysicist Fred Hoyle during a BBC radio broadcast, intended originally as a derogatory description of Lemaître's hypothesis. Yet, over time, it would come to symbolize the explosive birth of the universe itself.

1.3.1 The Hot, Dense Early Universe

According to the Big Bang model, the universe originated from a singular point of infinite density and temperature—a cosmic singularity. In the instant following the Big Bang, the universe underwent a period of rapid expansion, known as cosmic inflation, during which space itself stretched exponentially.

1.3.2 Formation of Matter and Light

As the universe cooled, elementary particles such as protons, neutrons, and electrons began to form, eventually coalescing into the first atoms. This epoch, known as recombination, marked the birth of light as photons decoupled from matter, filling the universe with a sea of radiation.

1.4 Echoes of the Big Bang

Though the intense radiation of the early universe has long since dissipated, echoes of the Big Bang persist in the form of the cosmic microwave background radiation (CMB). Discovered serendipitously in 1965 by Arno Penzias and Robert Wilson, the CMB represents the residual heat of the primordial fireball, serving as a window into the universe's infancy.

1.4.1 Cosmic Symphony: Anisotropies in the CMB

Subsequent observations of the CMB by missions such as the Cosmic Background Explorer (COBE) and the Wilkinson Microwave Anisotropy Probe (WMAP) have revealed subtle fluctuations or anisotropies in its temperature across the sky. These fluctuations provide crucial insights into the composition, geometry, and evolution of the universe.

1.5 Beyond the Big Bang

While the Big Bang theory provides a remarkably successful framework for understanding much of cosmic history, it is not without its unanswered questions and unresolved mysteries. From the nature of dark matter and dark energy to the ultimate fate of the universe itself, the quest to comprehend the cosmos continues unabated.

1.5.1 The Enigma of Dark Matter

One of the most pressing puzzles in modern cosmology is the nature of dark matter—a mysterious, invisible substance that exerts gravitational influence on galaxies and clusters of galaxies. Despite decades of research, its composition remains elusive, challenging our understanding of the fundamental constituents of the universe.

1.5.2 Dark Energy: Cosmic Acceleration

Equally perplexing is the phenomenon of dark energy, a mysterious force driving the accelerated expansion of the universe. First inferred from observations of distant supernovae in the late 1990s, dark energy constitutes the majority of the universe's energy density, yet its origin and properties remain shrouded in uncertainty.

1.6 Conclusion

As we reflect on the journey from the primordial singularity to the vast expanse of the cosmos we inhabit today, it becomes clear that the Big Bang theory represents not merely a scientific theory but a profound testament to the human capacity for understanding and exploration. With each discovery and each revelation, we inch closer to unlocking the secrets of the universe and charting our place within it.

Chapter 2: Early Universe: Primordial Soup

In the infancy of cosmic evolution, the universe was a seething cauldron of energy and matter, a dense and hot expanse where the seeds of structure and complexity were sown. This primordial epoch, characterized by extreme temperatures and densities, laid the foundation for the rich tapestry of cosmic phenomena that would follow. From the formation of elementary particles to the emergence of the first structures, the early universe was a crucible of transformation and innovation.

2.1 From Quarks to Quasars: The Particle Zoo

In the immediate aftermath of the Big Bang, the universe was a seething sea of energy, suffused with a primordial plasma of quarks, gluons, and leptons. During the first fraction of a second, known as the quark epoch, these elementary particles roiled and interacted in a frenzied dance, governed by the fundamental forces of nature.

2.1.1 The Standard Model of Particle Physics

The framework for understanding the behavior of particles and their interactions was laid out in the Standard Model of particle physics—a comprehensive theory that incorporates electromagnetism, the weak nuclear force, the strong nuclear force, and the Higgs mechanism. Through a combination of experimental observation and theoretical inference, scientists have constructed a remarkably successful description of the subatomic world.

2.1.2 The Era of Symmetry Breaking

As the universe cooled and expanded, it underwent a series of phase transitions, akin to freezing water into ice. During these transitions, symmetries that had previously held true were broken, leading to the emergence of distinct particle species and the formation of matter as we know it. The details of these symmetry-breaking processes are still the subject of intense study and debate.

2.2 The Formation of Nuclei: A Cosmic Alchemy

As the universe continued to cool, protons and neutrons began to combine to form atomic nuclei in a process known as nucleosynthesis. This epoch, occurring minutes after the Big Bang, marked the synthesis of the light elements hydrogen, helium, and trace amounts of lithium—a cosmic alchemy that set the stage for the formation of stars and galaxies.

2.2.1 Big Bang Nucleosynthesis

The conditions required for primordial nucleosynthesis were exquisitely sensitive, depending on the precise balance between the expansion rate of the universe and the strength of nuclear interactions. Through detailed calculations and comparisons with observational data, scientists have been able to constrain the parameters of Big Bang nucleosynthesis, providing crucial insights into the early universe.

2.3 The Cosmic Microwave Background: Fossil Light

A defining feature of the early universe is the cosmic microwave background radiation (CMB)—the residual glow of the hot, dense plasma that filled the cosmos in its infancy. Born from the decoupling of photons from matter roughly 380,000 years after the Big Bang, the CMB provides a snapshot of the universe at a time when it transitioned from opaque to transparent.

2.3.1 Mapping the Primordial Universe

Through meticulous observations by spacecraft such as the Cosmic Background Explorer

(COBE), the Wilkinson Microwave Anisotropy Probe (WMAP), and the Planck satellite, scientists have mapped the CMB with unprecedented precision, revealing subtle variations in temperature across the sky. These fluctuations encode valuable information about the composition, geometry, and history of the universe.

2.4 Inflationary Cosmology: The Quantum Genesis

While the Big Bang theory provides a robust framework for understanding much of cosmic evolution, it is not without its shortcomings. One of the most pressing challenges is the horizon problem—the apparent uniformity of the CMB on opposite sides of the sky, despite insufficient time for causal communication between them. Inflationary cosmology offers a compelling solution to this conundrum.

2.4.1 The Inflationary Epoch

Proposed in the early 1980s by physicists Alan Guth and Andrei Linde, inflation posits that the universe underwent a brief period of exponential expansion in the first fraction of a second, driven by a hypothetical scalar field known as the inflaton. This rapid expansion stretched quantum fluctuations to macroscopic scales, smoothing out the fabric of spacetime and providing a mechanism for generating the primordial density perturbations observed in the CMB.

2.4.2 Observational Signatures of Inflation

While inflation remains a theoretical construct, its predictions have found remarkable agreement with observational data. In particular, inflationary models predict a specific pattern of polarization in the CMB, known as B-mode polarization, which arises from primordial gravitational waves generated during the inflationary epoch. Detection of these elusive signals would provide smoking-gun evidence for the inflationary paradigm.

2.5 Conclusion

As we peer back through the mists of time to the tumultuous beginnings of the cosmos, the early universe emerges as a crucible of transformation and innovation. From the primordial soup of elementary particles to the emergence of the first structures, each epoch in cosmic history leaves its indelible mark on the fabric of spacetime. As we continue to probe the mysteries of the universe, we stand poised on the brink of discovery, ready to unlock the secrets of our cosmic origins.

Chapter 3: Formation of Galaxies and Stars

In the vast expanse of the cosmos, galaxies stand as cosmic islands, each harboring billions to trillions of stars bound together by gravity. From the majestic spirals to the enigmatic ellipticals, galaxies come in a variety of shapes and sizes, each a testament to the intricate dance of cosmic forces that govern their formation and evolution. In this chapter, we embark on a journey through the cosmic web, tracing the origins of galaxies and the birth of stars within their swirling depths.

3.1 The Cosmic Playground: Birth of Structure

The seeds of cosmic structure were sown in the primordial soup of the early universe, where tiny fluctuations in density provided the fertile ground for the growth of cosmic structures. Through the process of gravitational instability, these density perturbations gave rise to the vast cosmic web—a network of filaments, voids, and clusters that spans the universe.

3.1.1 From Quantum Fluctuations to Cosmic Structures

The origins of cosmic structure can be traced back to quantum fluctuations in the primordial universe, amplified by inflationary expansion and imprinted on the fabric of spacetime. Over billions of years, these fluctuations grew under the influence of gravity, eventually collapsing to form the first galaxies and galaxy clusters.

3.1.2 Simulations of Cosmic Evolution

To unravel the complex interplay of gravitational dynamics, gas physics, and star formation that governs the formation of galaxies, astrophysicists employ sophisticated numerical simulations. By modeling the evolution of billions of particles within a virtual universe, these simulations provide valuable insights into the emergence of cosmic structure and the distribution of galaxies on large scales.

3.2 Galactic Collisions and Mergers: Cosmic Ballet

Galaxies are not static entities but dynamic systems that interact and evolve over cosmic time. One of the most dramatic events in galactic evolution is the collision and merger of galaxies—a cosmic ballet that shapes their morphology, triggers star formation, and drives the growth of supermassive black holes at their centers.

3.2.1 The Dance of Galaxies

Galactic collisions and mergers can take on a variety of forms, from gentle encounters between dwarf galaxies to violent collisions between massive spirals. As galaxies approach each other, gravitational forces distort their shapes, triggering bursts of star formation and feeding their central black holes with gas and dust.

3.2.2 Observational Evidence

Observations of interacting galaxies in the local and distant universe provide compelling evidence for the role of mergers in galactic evolution. Through techniques such as deep imaging, spectroscopy, and radio observations, astronomers can trace the signatures of past collisions and mergers, revealing the complex interplay of stars, gas, and dark matter.

3.3 Star Formation: Cosmic Crucibles

At the heart of every galaxy lies a stellar nursery—a region of intense gas and dust where stars are born in the crucible of gravity and pressure. From the luminous giants to the dim dwarfs, stars come in a variety of sizes and colors, each with its own story of formation and evolution.

3.3.1 Protostellar Collapse

The journey to star formation begins with the gravitational collapse of a dense molecular cloud—a vast reservoir of gas and dust scattered throughout the galaxy. As the cloud contracts under its own gravity, it fragments into smaller clumps, each destined to become a new generation of stars.

3.3.2 The Birth of a Star

Within these clumps, known as protostellar cores, dense knots of gas and dust coalesce to form protostars—precursors to fully-fledged stars. As the protostar accretes material from its surrounding envelope, it grows increasingly luminous, eventually igniting nuclear fusion in its core and shining as a bona fide star.

3.4 Stellar Evolution: Life and Death of Stars

Once ignited, stars undergo a remarkable journey through the cosmic depths, transforming over millions to billions of years as they fuse hydrogen into heavier elements and evolve through various stages of stellar evolution. From the fiery birth of massive stars to the serene demise of white dwarfs, the life cycles of stars are as diverse as they are awe-inspiring.

3.4.1 Main Sequence Stars

The majority of stars, including our own Sun, spend the majority of their lives on the main sequence—a stable phase of hydrogen fusion in their cores. During this stage, the outward pressure of nuclear fusion balances the inward pull of gravity, maintaining a delicate equilibrium that sustains the star for billions of years.

3.4.2 Stellar Death and Renewal

As stars exhaust their hydrogen fuel, they undergo a series of transformations, culminating in their eventual demise. Low to medium-mass stars, like the Sun, will eventually shed their outer layers to form planetary nebulae, leaving behind a dense core known as a white dwarf. In contrast, massive stars will end their lives in cataclysmic supernova explosions, seeding the cosmos with heavy elements and, in some cases, collapsing to form neutron stars or black holes.

3.5 Conclusion

From the primordial fluctuations of the early universe to the fiery furnaces of stellar nurseries, the formation and evolution of galaxies and stars are a testament to the intricate interplay of cosmic forces. As we continue to probe the depths of the cosmos, each new discovery offers a glimpse into the rich tapestry of cosmic history, revealing the remarkable diversity and complexity of the universe we inhabit.

Chapter 4: Stellar Evolution: Birth, Life, and Death of Stars

Stars are the celestial engines that power the universe, transforming matter into energy through the process of nuclear fusion. From their fiery births in stellar nurseries to their serene deaths as white dwarfs, neutron stars, or black holes, stars follow a remarkable journey through cosmic time. In this chapter, we delve into the intricacies of stellar evolution, exploring the diverse pathways that stars traverse as they illuminate the cosmic landscape.

4.1 The Birth of Stars: From Cosmic Clouds to Protostars

The journey to stellar birth begins in the cold, dark recesses of molecular clouds—vast reservoirs of gas and dust scattered throughout the galaxy. Within these clouds, gravitational instabilities trigger the collapse of dense regions, leading to the formation of protostellar cores—precursors to fully-fledged stars.

4.1.1 Gravitational Collapse and Protostellar Formation

As a molecular cloud collapses under its own gravity, it fragments into smaller clumps, each destined to become a new generation of stars. Within these clumps, known as protostellar cores, gas and dust accrete onto a central protostar, heating it up and initiating nuclear fusion in its core.

4.1.2 Stellar Nurseries: The Cosmic Cradles of Star Formation

Stellar nurseries are regions of intense gas and dust where star formation is most prolific. These nurseries are often associated with giant molecular clouds, where the interplay of gravity, magnetic fields, and turbulence creates ideal conditions for the birth of stars. Examples of stellar nurseries include the Orion Nebula and the Eagle Nebula, where thousands of young stars are actively forming.

4.2 Main Sequence Stars: The Stellar Engines of the Cosmos

Once ignited, stars enter a stable phase of hydrogen fusion known as the main sequence, where they spend the majority of their lives. During this phase, the outward pressure of nuclear fusion balances the inward pull of gravity, maintaining a delicate equilibrium that sustains the star for billions of years.

4.2.1 The Hertzsprung-Russell Diagram

The Hertzsprung-Russell (H-R) diagram is a powerful tool used by astronomers to classify and study stars based on their luminosity, temperature, and evolutionary stage. By plotting stars on a graph with luminosity on the vertical axis and temperature or spectral type on the horizontal axis, astronomers can identify different stages of stellar evolution, from protostars to white dwarfs.

4.2.2 The Sun: A Typical Main Sequence Star

Our own Sun is a middle-aged, main sequence star of spectral type G2V, with a surface temperature of approximately 5,500 degrees Celsius. It is currently fusing hydrogen into helium in its core, a process that releases energy and sustains the Sun's luminosity. The Sun is expected to remain on the main sequence for another 5 billion years before evolving into a red giant.

4.3 Stellar Death: The Final Act

As stars exhaust their nuclear fuel, they undergo a series of transformations that ultimately lead to their demise. The fate of a star depends on its initial mass: low to medium-mass stars like the Sun will end their lives as white dwarfs, while massive stars will undergo supernova explosions and may collapse to form neutron stars or black holes.

4.3.1 The Death of Low to Medium-Mass Stars

When a low to medium-mass star like the Sun exhausts its nuclear fuel, it undergoes a series of transformations that culminate in the formation of a planetary nebula and a white dwarf. During the red giant phase, the star expands and sheds its outer layers, exposing its hot core. The remaining core, composed of degenerate matter, cools over billions of years to form a white dwarf—a dense, Earth-sized remnant of the star's former self.

4.3.2 Supernova Explosions and Stellar Remnants

In contrast, massive stars with initial masses greater than eight times that of the Sun will undergo supernova explosions at the end of their lives. During a supernova, the outer layers of the star are ejected into space at high velocities, leaving behind a compact remnant such as a neutron star or black hole. These exotic objects represent the endpoints of stellar evolution, where the forces of gravity and nuclear physics push the boundaries of our understanding.

4.4 Stellar Feedback: The Impact of Stars on their Environment

Throughout their lives, stars exert a profound influence on their surroundings through a process known as stellar feedback. From the ionizing radiation of hot, young stars to the shockwaves generated by supernova explosions, stellar feedback plays a crucial role in shaping the structure and evolution of galaxies and the interstellar medium.

4.4.1 Ionizing Radiation and HII Regions

Massive, hot stars emit copious amounts of ultraviolet radiation that ionizes the surrounding gas, creating glowing regions known as HII regions. These regions serve as cosmic signposts of ongoing star formation and provide valuable insights into the properties of young, massive stars.

4.4.2 Supernova Explosions and Galactic Evolution

Supernova explosions inject vast amounts of energy and heavy elements into the interstellar medium, triggering shockwaves that sweep through the galaxy and trigger subsequent generations of star formation. This process, known as supernova feedback, plays a crucial role in regulating the rate of star formation and shaping the chemical composition of galaxies over cosmic time.

4.5 Conclusion

As we reflect on the journey from stellar birth to stellar death, it becomes clear that stars are not merely celestial objects but cosmic engines that drive the evolution of the universe. From the fiery furnaces of stellar nurseries to the serene glow of white dwarfs, each stage in the life cycle of a star offers a glimpse into the rich tapestry of cosmic history, revealing the intricate interplay of gravity, nuclear physics, and stellar feedback that governs the cosmos. As we continue to explore the mysteries of the universe, the story of stellar evolution serves as a reminder of the beauty and complexity of the cosmos we inhabit.

Chapter 5: Black Holes: Nature's Cosmic Vacuum Cleaners

In the vast expanse of the universe, black holes stand as some of the most enigmatic and mysterious objects known to science. Born from the remnants of massive stars or the collapse of dense regions in the early universe, black holes possess gravitational fields so intense that not even light can escape their grasp. In this chapter, we embark on a journey into the heart of darkness, exploring the properties, formation, and behavior of these cosmic vacuum cleaners.

5.1 The Concept of Black Holes: A Brief History

The idea of black holes has its roots in the work of early pioneers of theoretical physics, including Albert Einstein and Karl Schwarzschild. However, it was not until the 20th century that the concept began to take shape, thanks to the groundbreaking work of scientists such as Subrahmanyan Chandrasekhar and John Wheeler.

5.1.1 Einstein's Theory of General Relativity

The foundation for our modern understanding of black holes lies in Einstein's theory of general relativity, published in 1915. According to general relativity, gravity is not simply a force between objects but rather the curvature of spacetime caused by the presence of mass and energy. Black holes emerge as one of the most striking consequences of this geometric description of gravity.

5.1.2 Schwarzschild's Solution

In 1916, Karl Schwarzschild, a German physicist serving on the Eastern Front during World War I, derived the first exact solution to Einstein's field equations—a description of the spacetime geometry surrounding a non-rotating, spherically symmetric mass. This solution, now known as the Schwarzschild metric, laid the groundwork for understanding the properties of black holes.

5.2 Anatomy of a Black Hole: Event Horizons and Singularities

At the heart of every black hole lies a region of spacetime known as the singularity—a point of infinite density and curvature where the laws of physics break down. Surrounding the singularity is the event horizon—the boundary beyond which nothing, not even light, can escape the gravitational pull of the black hole.

5.2.1 Event Horizons and the Point of No Return

The event horizon of a black hole marks the boundary beyond which escape is impossible—a point of no return for anything that ventures too close. Once an object crosses the event horizon, it is inexorably drawn into the black hole's gravitational grip, destined to be consumed by the singularity at its center.

5.2.2 Types of Black Holes

Black holes come in a variety of flavors, classified primarily based on their mass and angular momentum. Stellar-mass black holes, formed from the remnants of massive stars, typically have masses ranging from a few times that of the Sun to tens of times the Sun's mass. Supermassive black holes, found at the centers of galaxies, can have masses millions to billions of times greater than that of the Sun.

5.3 Formation of Black Holes: Cosmic Cataclysms

Black holes can form through a variety of processes, including the collapse of massive stars, the mergers of compact objects, and the collapse of dense regions in the early universe. Each of these pathways gives rise to black holes with different properties and observational signatures.

5.3.1 Stellar-Mass Black Holes

Stellar-mass black holes form when massive stars exhaust their nuclear fuel and undergo core collapse, leading to a catastrophic supernova explosion. If the core remnant is sufficiently massive, it may collapse to form a black hole, where gravitational forces overwhelm all other forces and compress the matter into a singularity.

5.3.2 Supermassive Black Holes

The origins of supermassive black holes, found at the centers of most galaxies, remain a subject of active research and debate. One leading hypothesis is that they form through the accretion of gas and stellar debris onto a central seed black hole, which grows over time through mergers with other black holes and the accretion of surrounding material.

5.4 Black Hole Astrophysics: Jets, Accretion Disks, and Quasars

Black holes are not just cosmic vacuum cleaners but also powerful engines that drive some of the most energetic phenomena in the universe. From the relativistic jets emitted by supermassive black holes to the luminous accretion disks that surround them, black hole astrophysics offers a window into the extreme physics of the cosmos.

5.4.1 Accretion Disks: Cosmic Powerhouses

When matter falls into a black hole's gravitational well, it forms a swirling disk of gas and dust known as an accretion disk. As material spirals inward, it releases gravitational potential energy in the form of heat and radiation, producing intense X-ray and gamma-ray emission that can be detected by telescopes on Earth and in space.

5.4.2 Relativistic Jets: Cosmic Cannons

In some cases, black holes can launch highly collimated jets of plasma at near-light speeds, extending for millions of light-years across the cosmos. These relativistic jets are thought to arise from the twisting and shearing of magnetic fields near the black hole's event horizon, where powerful gravitational forces generate enormous amounts of energy.

5.5 Black Holes and Galactic Evolution

Black holes play a crucial role in shaping the evolution of galaxies and the cosmic landscape. Through processes such as accretion, mergers, and feedback, black holes influence the formation of stars, the growth of galaxies, and the distribution of matter on large scales.

5.5.1 Galaxy-Black Hole Coevolution

Observations suggest a close relationship between the growth of supermassive black holes and the evolution of their host galaxies—a phenomenon known as galaxy-black hole coevolution. As galaxies merge and accrete gas, they fuel the growth of their central black holes, which in turn release energy and influence the surrounding environment through processes such as quasar activity and jet formation.

5.5.2 Quasars: Beacons of Cosmic Evolution

Quasars are among the most luminous objects in the universe, powered by the accretion of material onto supermassive black holes at the centers of galaxies. These cosmic beacons shine brightly across vast distances, illuminating the early universe and providing valuable insights into the growth of galaxies and the formation of cosmic structures.

5.6 Conclusion

As we peer into the depths of space and time, black holes emerge as some of the most fascinating and enigmatic objects in the universe. From their exotic properties to their profound influence on cosmic evolution, black holes challenge our understanding of gravity, spacetime, and the fundamental nature of the cosmos. As we continue to explore the mysteries of the universe, black holes stand as cosmic gateways to the unknown, beckoning us to unravel their secrets and unlock the mysteries of the cosmos.

Chapter 6: Exoplanets: Worlds Beyond Our Solar System

In the grand tapestry of the cosmos, planets have long been regarded as the jewels of the universe, providing a stage for the drama of life and evolution. While our own solar system boasts a diverse array of planets, moons, and other celestial bodies, the discovery of exoplanets—worlds orbiting distant stars—has revolutionized our understanding of planetary systems and the potential for life beyond Earth. In this chapter, we embark on a journey to explore the rich diversity of exoplanets and the quest to uncover the secrets of alien worlds.

6.1 A Brief History of Exoplanet Discovery

The quest to find exoplanets dates back centuries, with astronomers and philosophers speculating about the existence of other worlds beyond our own. However, it was not until the late 20th century that technological advancements enabled the first confirmed detections of exoplanets orbiting nearby stars.

6.1.1 The Radial Velocity Method

One of the earliest methods for detecting exoplanets is the radial velocity or Doppler method, which relies on the gravitational tug exerted by an orbiting planet on its parent star. As a planet orbits its star, it causes the star to wobble slightly, inducing periodic shifts in its spectral lines that can be detected using spectroscopic techniques.

6.1.2 The Transit Method

Another powerful technique for discovering exoplanets is the transit method, which relies on the periodic dimming of a star's light as a planet passes in front of it, or transits, from our line of sight. By measuring the slight decrease in brightness during these transits, astronomers can infer the presence and properties of orbiting planets.

6.2 The Diversity of Exoplanets

The discovery of exoplanets has revealed a staggering diversity of worlds, ranging from hot Jupiters and super-Earths to icy giants and Earth-like planets. Each of these exoplanets offers a unique window into the processes of planetary formation and evolution, challenging our preconceptions about the nature of planetary systems.

6.2.1 Hot Jupiters and Super-Earths

Some of the earliest exoplanets discovered were hot Jupiters—gas giants that orbit close to their parent stars in tight, blistering orbits. These worlds defy the expectations of traditional planetary formation theories, raising questions about their origins and migration mechanisms. In contrast, super-Earths are rocky planets with masses greater than that of Earth but less than that of Neptune, offering insights into the diversity of terrestrial worlds in the universe.

6.2.2 Habitable Zone Planets

One of the most tantalizing discoveries in exoplanet research is the identification of planets orbiting within the habitable zone of their parent stars—the region where conditions may be conducive to the existence of liquid water and, potentially, life as we know it. While the search for true Earth analogs remains ongoing, the discovery of habitable zone planets represents a significant step forward in our quest to find life beyond Earth.

6.3 Planetary Formation and Evolution

The study of exoplanets provides valuable insights into the processes of planetary formation and evolution, shedding light on the conditions that give rise to diverse planetary systems and the factors that shape their long-term evolution.

6.3.1 The Core Accretion Model

One leading theory of planetary formation is the core accretion model, which posits that planets form from the gradual accumulation of solid material within a protoplanetary disk surrounding a young star. Over time, dust grains collide and coalesce to form larger bodies known as planetesimals, which eventually grow into planets through the process of accretion and gravitational collapse.

6.3.2 Planetary Migration

Recent studies suggest that many exoplanets may undergo significant migration during their early history, moving inward or outward from their original orbits due to interactions with their parent star or other planets in the system. This planetary migration can have profound implications for the architecture and stability of planetary systems, leading to the formation of hot Jupiters, eccentric orbits, and resonant configurations.

6.4 The Search for Extraterrestrial Life

One of the most profound questions in exoplanet research is whether these distant worlds harbor life—or even civilizations—beyond Earth. While the discovery of habitable zone planets raises hopes for finding signs of life elsewhere in the universe, the search for extraterrestrial intelligence remains a daunting challenge that requires interdisciplinary collaboration and technological innovation.

6.4.1 Biosignatures and Atmospheric Analysis

To search for signs of life on exoplanets, astronomers study their atmospheres for the presence of certain molecules or combinations of molecules known as biosignatures. These could include oxygen, methane, or other gases that are indicative of biological activity. By analyzing the composition and chemistry of exoplanet atmospheres, scientists hope to identify potential habitats for life beyond Earth.

6.4.2 The Drake Equation and the Fermi Paradox

The search for extraterrestrial intelligence is guided by theoretical frameworks such as the Drake equation, which estimates the number of technologically advanced civilizations in the Milky Way galaxy based on factors such as the rate of star formation, the fraction of stars with planets, and the probability of life emerging on a habitable planet. However, the absence of direct evidence for extraterrestrial civilizations, known as the Fermi paradox, underscores the challenges and uncertainties inherent in the search for intelligent life in the universe.

6.5 The Future of Exoplanet Exploration

As technology continues to advance, the search for exoplanets and the quest for understanding their properties and potential for life will only intensify. From space-based observatories to ground-based telescopes, a new generation of instruments and missions is poised to revolutionize our understanding of exoplanets and their place in the cosmos.

6.5.1 Space-Based Missions: Kepler, TESS, and Beyond

The Kepler Space Telescope, launched by NASA in 2009, revolutionized the field of exoplanet research by surveying a patch of sky in the constellation Cygnus and detecting thousands of exoplanet candidates using the transit method. Although Kepler's primary mission ended in 2013, its legacy lives on through the continued analysis of its data and the discovery of exoplanets within its field of view.

Building upon Kepler's success, NASA's Transiting Exoplanet Survey Satellite (TESS), launched in 2018, aims to expand the search for exoplanets to the entire sky. By scanning nearby stars for the telltale dips in brightness caused by transiting planets, TESS is expected to discover thousands of new exoplanets, including potentially habitable worlds ripe for follow-up study.

Looking ahead, future space-based missions such as the James Webb Space Telescope (JWST) and the PLATO mission by the European Space Agency (ESA) promise to further revolutionize our understanding of exoplanets. JWST, with its unprecedented sensitivity and infrared capabilities, will enable detailed observations of exoplanet atmospheres and the search for signs of life beyond Earth. PLATO, on the other hand, will focus on characterizing rocky planets around nearby stars, with the goal of identifying potentially habitable worlds.

6.5.2 Ground-Based Observatories and Next-Generation Instruments

In addition to space-based missions, ground-based observatories and next-generation instruments are playing a crucial role in the study of exoplanets. Telescopes such as the Very

Large Telescope (VLT) in Chile and the Keck Observatory in Hawaii are equipped with advanced instrumentation for studying exoplanet atmospheres, detecting exoplanet-hosting binary star systems, and characterizing the properties of distant worlds.

Furthermore, upcoming ground-based projects such as the Giant Magellan Telescope (GMT) and the Extremely Large Telescope (ELT) are poised to push the boundaries of exoplanet research even further. With their unprecedented size and sensitivity, these next-generation telescopes will enable astronomers to study exoplanets with unparalleled precision, paving the way for groundbreaking discoveries in the years to come.

6.6 Ethical and Philosophical Implications

The search for exoplanets and the quest for understanding life beyond Earth raise profound ethical and philosophical questions about humanity's place in the universe and our responsibilities as stewards of life on Earth. As we explore distant worlds and contemplate the possibility of extraterrestrial life, it is essential to consider the potential impact of our discoveries on society, culture, and the environment.

6.6.1 Planetary Protection and Interstellar Contamination

One of the key ethical considerations in exoplanet exploration is the risk of interstellar contamination—the inadvertent transfer of Earthly microbes to other planets or vice versa. To mitigate this risk, space agencies and international organizations have developed strict planetary protection protocols to minimize the likelihood of biological contamination during space missions.

6.6.2 The Search for Intelligent Life

The search for extraterrestrial intelligence (SETI) raises complex questions about the nature of consciousness, communication, and the potential for interstellar contact. While the discovery of intelligent life beyond Earth would undoubtedly be a profound and transformative event, it also raises ethical dilemmas regarding our responsibilities as stewards of the cosmos and the potential consequences of contact with advanced civilizations.

6.7 Conclusion

As we peer into the depths of space and time, the discovery of exoplanets opens a window onto the vast diversity of worlds beyond our solar system. From scorching gas giants to rocky terrestrial planets, each exoplanet offers a unique glimpse into the rich tapestry of cosmic evolution and the potential for life beyond Earth.

As we continue to explore the mysteries of the universe, the study of exoplanets promises to revolutionize our understanding of planetary systems, galactic evolution, and the fundamental nature of the cosmos. Through interdisciplinary collaboration, technological innovation, and ethical reflection, we stand poised on the brink of a new era of exploration—one that will shape our understanding of the universe and our place within it for generations to come.

Chapter 7: Dark Matter: Unseen Influences in the Cosmos

In the vast expanse of the cosmos, there exists a mysterious substance that defies detection by conventional means yet exerts a powerful gravitational influence on the structures we observe. This enigmatic substance, known as dark matter, constitutes the majority of matter in the universe and plays a crucial role in shaping the formation and evolution of galaxies and cosmic structures. In this chapter, we delve into the mysteries of dark matter, exploring its properties, its role in the universe, and the ongoing efforts to unlock its secrets.

7.1 The Quest for Dark Matter: A Brief History

The concept of dark matter emerged from the pioneering work of astronomers and physicists in the early 20th century who sought to understand the distribution of matter in the universe and the dynamics of galaxies and galaxy clusters.

7.1.1 Fritz Zwicky and Galaxy Dynamics

In the 1930s, Swiss astronomer Fritz Zwicky made a groundbreaking discovery while studying the Coma Cluster of galaxies. By measuring the velocities of galaxies within the cluster, Zwicky found that their motions far exceeded what could be accounted for by the visible matter alone. He hypothesized the existence of unseen, or "dark," matter to explain this gravitational anomaly—a concept that would later become central to our understanding of the cosmos.

7.1.2 Vera Rubin and Galactic Rotation Curves

In the 1970s, astronomer Vera Rubin and her colleague Kent Ford conducted pioneering observations of the rotation curves of spiral galaxies. To their surprise, they found that stars and gas in the outer regions of galaxies were moving at velocities that defied the predictions of Newtonian gravity based on the visible mass alone. Their work provided further evidence for the presence of dark matter in galaxies, revolutionizing our understanding of galactic dynamics.

7.2 What is Dark Matter?

Despite decades of research, the true nature of dark matter remains one of the greatest unsolved mysteries in physics. While its properties are still largely unknown, scientists have developed several compelling hypotheses to explain its existence and behavior.

7.2.1 Weakly Interacting Massive Particles (WIMPs)

One leading candidate for dark matter is the weakly interacting massive particle, or WIMP. According to this hypothesis, dark matter consists of elusive particles that interact only through the weak nuclear force and gravity, making them extremely difficult to detect using conventional means. While numerous experiments have been conducted to search for WIMPs, none have yielded conclusive evidence thus far.

7.2.2 Axions and Axion-Like Particles

Another intriguing possibility is that dark matter may be composed of axions or axion-like particles—hypothetical particles that arise from extensions of the standard model of particle physics. Axions are predicted to have very low masses and interact extremely weakly with ordinary matter, making them challenging to detect experimentally. Nevertheless, ongoing experiments aim to test the axion hypothesis and shed light on the nature of dark matter.

7.3 Observational Evidence for Dark Matter

While dark matter itself remains invisible to telescopes and other astronomical instruments, its presence is inferred from its gravitational effects on visible matter and light. Over the years, astronomers have amassed a wealth of observational evidence for the existence of dark matter, providing strong support for its role as the dominant form of matter in the universe.

7.3.1 Gravitational Lensing

One of the most compelling lines of evidence for dark matter comes from the phenomenon of gravitational lensing. As light from distant galaxies travels through the universe, it can be bent and distorted by the gravitational influence of massive objects such as galaxy clusters. By studying the patterns of gravitational lensing, astronomers can map the distribution of dark matter in the cosmos and infer its properties.

7.3.2 Cosmic Microwave Background

Another crucial piece of evidence for dark matter comes from observations of the cosmic microwave background (CMB)—the faint afterglow of the Big Bang. Tiny fluctuations in the temperature and polarization of the CMB provide valuable clues about the composition and distribution of matter in the early universe, including dark matter. By analyzing these fluctuations, scientists can constrain the amount of dark matter present in the cosmos and its influence on cosmic structure formation.

7.4 The Role of Dark Matter in Galactic and Cosmic Evolution

Dark matter plays a central role in shaping the formation and evolution of galaxies and cosmic structures on large scales. From the gravitational collapse of primordial fluctuations to the assembly of galaxy clusters, dark matter acts as the cosmic scaffolding upon which visible matter is assembled and organized.

7.4.1 Galaxy Formation and Dynamics

In the hierarchical model of galaxy formation, small fluctuations in the density of dark matter in the early universe serve as seeds for the formation of cosmic structures. As dark matter halos grow through gravitational collapse, they attract and accrete gas and dust, eventually forming galaxies like the Milky Way. The distribution and dynamics of dark matter within galaxies influence the motions of stars and gas, shaping the observed properties of galaxies and their rotation curves.

7.4.2 Large-Scale Structure Formation

On even larger scales, dark matter drives the formation of cosmic web—a vast network of filaments and voids that permeates the universe. Galaxies and galaxy clusters are found to be distributed along these cosmic filaments, with dark matter acting as the gravitational glue that binds them together. By studying the large-scale distribution of galaxies and the cosmic web, astronomers can infer the properties of dark matter and its role in cosmic evolution.

7.5 The Search for Dark Matter Particles

Despite decades of research, dark matter particles have yet to be directly detected in laboratory experiments. Nevertheless, scientists continue to search for elusive dark matter particles using a variety of experimental techniques and approaches.

7.5.1 Direct Detection Experiments

One approach to detecting dark matter particles is to search for their interactions with ordinary matter in underground laboratory experiments. These experiments typically involve large detectors located deep underground to shield them from cosmic rays and other sources of background radiation. If a dark matter particle collides with an atomic nucleus in the detector, it may produce a detectable signal that can be recorded and analyzed.

7.5.2 Indirect Detection Experiments

Another approach is to search for the products of dark matter annihilation or decay in cosmic-ray observations and astrophysical phenomena. Dark matter particles may annihilate or decay into other particles, such as photons, neutrinos, or charged particles, which can be detected by telescopes and detectors on Earth and in space. By searching for these indirect signatures of dark matter, scientists hope to uncover clues about its properties and interactions.

7.6 Alternative Theories of Gravity

While the existence of dark matter provides a compelling explanation for a wide range of astronomical phenomena, some scientists have proposed alternative theories of gravity that seek to modify or extend Einstein's theory of general relativity. These theories, which include modified Newtonian dynamics (MOND) and modified gravity models, aim to explain the observed dynamics of galaxies and galaxy clusters without the need for dark matter.

7.6.1 Modified Newtonian Dynamics (MOND)

MOND is a modified theory of gravity proposed by Israeli physicist Mordehai Milgrom in the 1980s as an alternative to dark matter. According to MOND, the gravitational force experienced by stars and galaxies deviates from the predictions of Newtonian gravity at very low accelerations, leading to modified dynamics in the outer regions of galaxies. While MOND has had some success in explaining certain observed phenomena, such as galactic rotation curves, it has struggled to account for the full range of evidence supporting the existence of dark matter.

7.6.2 Modified Gravity Models

In addition to MOND, various modified gravity models have been proposed to explain the observed dynamics of galaxies and galaxy clusters without invoking dark matter. These models typically involve modifications to the equations of general relativity or the introduction of additional fields or dimensions of spacetime. While these theories offer intriguing alternatives to dark matter, they have yet to be conclusively supported by observational evidence.

7.7 Unresolved Questions and Future Directions

Despite significant progress in our understanding of dark matter, many questions remain unanswered, and the search for dark matter particles continues to be one of the most active areas of research in astrophysics and particle physics.

7.7.1 Nature of Dark Matter Particles

One of the key outstanding questions is the nature of dark matter particles themselves. While numerous candidates have been proposed, ranging from WIMPs to axions, none have been conclusively detected. Future experiments, such as the Large Underground Xenon (LUX) experiment and the Axion Dark Matter eXperiment (ADMX), aim to push the limits of sensitivity and probe new regions of parameter space in the search for dark matter particles.

7.7.2 Distribution and Properties of Dark Matter

Another area of active research is the distribution and properties of dark matter on galactic and cosmic scales. By combining observations from a variety of sources, including gravitational lensing, galaxy dynamics, and large-scale structure surveys, astronomers hope to map the distribution of dark matter in unprecedented detail and constrain its properties, such as its mass and interaction cross-section.

7.7.3 Dark Matter in the Early Universe

Understanding the role of dark matter in the early universe is essential for constraining its properties and origins. By studying the imprint of dark matter on the cosmic microwave background and the large-scale structure of the universe, scientists can infer its abundance and properties in the early stages of cosmic evolution, shedding light on the processes that gave rise to dark matter in the first place.

7.8 Ethical and Societal Implications

The quest to understand dark matter and its role in the universe raises important ethical and societal questions about the nature of scientific inquiry, the allocation of resources, and the communication of scientific findings to the public.

7.8.1 Funding and Resource Allocation

Given the fundamental importance of dark matter to our understanding of the cosmos, it is essential to prioritize research and funding for experiments and observatories dedicated to its study. Ensuring continued support for dark matter research will require collaboration between governments, funding agencies, and the scientific community to address some of the most pressing questions in astrophysics and particle physics.

7.8.2 Public Engagement and Education

Communicating the significance of dark matter research to the public is crucial for fostering interest and understanding in science and astronomy. By engaging with the public through outreach events, educational programs, and media outreach, scientists can inspire future generations of researchers and help build a more scientifically literate society.

7.9 Conclusion

Dark matter stands as one of the most profound and enduring mysteries in the cosmos, challenging our understanding of the universe and the fundamental laws of physics. From its elusive nature to its far-reaching gravitational influence, dark matter continues to captivate the imagination of scientists and the public alike.

As we continue to probe the mysteries of the cosmos, the quest to understand dark matter remains one of the greatest scientific endeavors of our time. Through interdisciplinary collaboration, technological innovation, and a relentless spirit of inquiry, scientists are poised to unlock the secrets of dark matter and illuminate the hidden depths of the universe. In doing so, we may uncover new insights into the nature of matter, gravity, and the cosmos, reshaping our understanding of the universe and our place within it.

Chapter 8: Dark Energy: Mysterious Forces Shaping the Universe

In the grand theater of the cosmos, there exists a force that defies conventional understanding—a mysterious energy that permeates the vast expanses of space and drives the accelerated expansion of the universe. This enigmatic entity, known as dark energy, stands as one of the greatest puzzles in modern cosmology, challenging our notions of gravity, matter, and the fate of the cosmos. In this chapter, we embark on a journey to explore the mysteries of dark energy, from its discovery to its profound implications for the future of the universe.

8.1 The Discovery of Dark Energy: A Cosmic Surprise

The existence of dark energy was first inferred from observations of distant supernovae in the late 1990s, which revealed an unexpected phenomenon—the accelerated expansion of the universe. This groundbreaking discovery, made by two independent research teams led by Saul Perlmutter, Brian Schmidt, and Adam Riess, revolutionized our understanding of the cosmos and earned them the Nobel Prize in Physics in 2011.

8.1.1 Observational Evidence

The discovery of dark energy was based on observations of Type Ia supernovae—powerful explosions that occur when white dwarf stars in binary systems reach a critical mass and explode. By measuring the brightness and redshift of these supernovae, astronomers found that distant supernovae appeared fainter than expected, indicating that the universe is expanding at an accelerating rate.

8.1.2 Cosmic Microwave Background

In addition to supernova observations, measurements of the cosmic microwave background (CMB)—the afterglow of the Big Bang—provide valuable insights into the nature of dark energy. Variations in the temperature and polarization of the CMB reveal the large-scale structure of the universe and the distribution of matter and energy, including dark energy.

8.2 What is Dark Energy?

Dark energy is a mysterious form of energy that permeates the fabric of spacetime and exerts a repulsive gravitational force, driving the accelerated expansion of the universe. While its exact nature remains unknown, dark energy is thought to make up approximately 70% of the total energy density of the universe, making it the dominant component of the cosmos.

8.2.1 Cosmological Constant

The simplest explanation for dark energy is the cosmological constant—a constant energy density that fills empty space and drives the expansion of the universe. Proposed by Albert Einstein in his theory of general relativity, the cosmological constant acts as a repulsive force that counteracts the gravitational pull of matter, causing the universe to expand at an accelerating rate.

8.2.2 Quintessence and Dynamical Dark Energy

In addition to the cosmological constant, alternative models of dark energy have been proposed, such as quintessence and dynamical dark energy. These models posit that dark energy may vary over time or space, leading to changes in the rate of cosmic expansion. While these models offer intriguing alternatives to the cosmological constant, they have yet to be conclusively supported by observational evidence.

8.3 The Nature of Cosmic Acceleration

The discovery of cosmic acceleration poses profound questions about the nature of gravity, the fate of the universe, and the underlying structure of spacetime. Understanding the mechanisms driving cosmic acceleration is essential for unraveling the mysteries of dark energy and its implications for the cosmos.

8.3.1 Einstein's Equations of General Relativity

At the heart of our understanding of cosmic acceleration lies Einstein's theory of general relativity, which describes how matter and energy warp the fabric of spacetime and dictate the dynamics of the universe. According to general relativity, the presence of dark energy—whether in the form of a cosmological constant or dynamical field—leads to a repulsive gravitational force that drives the accelerated expansion of the universe.

8.3.2 Modified Gravity Theories

In addition to general relativity, alternative theories of gravity have been proposed to explain cosmic acceleration without invoking dark energy. These theories, which include modifications to Einstein's equations or the introduction of new fields or dimensions of spacetime, aim to reproduce the observed dynamics of the universe while avoiding the need for exotic forms of energy. While these theories offer intriguing alternatives to dark energy, they have yet to be conclusively supported by observational evidence.

8.4 Implications for the Fate of the Universe

The discovery of dark energy has profound implications for the ultimate fate of the universe, shaping our understanding of its past, present, and future evolution. Depending on the nature of dark energy, the universe may continue to expand indefinitely, experiencing a "big rip" in which galaxies, stars, and even atoms are torn apart by the relentless expansion of spacetime.

8.4.1 Open, Flat, and Closed Universes

The fate of the universe is determined by its overall geometry, which is governed by the balance between the expansion rate and the gravitational pull of matter and energy. In an open universe, the expansion rate exceeds the critical density, leading to infinite expansion and eventual heat death. In a closed universe, the expansion rate is insufficient to overcome gravity, leading to eventual collapse in a "big crunch." In a flat universe, the expansion rate precisely balances the critical density, leading to indefinite expansion and a stable cosmic future.

8.4.2 The Big Rip and Cosmic Doomsday

In some scenarios, dark energy may lead to a "big rip" in which the repulsive force of dark energy overwhelms the gravitational pull of matter, causing the universe to expand at an ever-increasing rate. In this scenario, galaxies, stars, and even atoms would be torn apart by the relentless expansion of spacetime, culminating in a cosmic doomsday billions of years in the future.

8.5 Theoretical Challenges and Future Directions

Despite significant progress in our understanding of dark energy, many questions remain unanswered, and the search for its true nature continues to be one of the most active areas of research in cosmology and theoretical physics.

8.5.1 The Nature of Dark Energy

One of the key challenges in dark energy research is determining the nature of dark energy itself. Is dark energy a cosmological constant, a dynamical field, or something else entirely? Answering this question requires a combination of observational data, theoretical modeling, and experimental testing to constrain the properties of dark energy and its effects on the universe.

8.5.2 Testing Alternative Theories of Gravity

In addition to dark energy, alternative theories of gravity offer intriguing possibilities for explaining cosmic acceleration. By testing these theories against observational data and experimental results, scientists can probe the fundamental nature of gravity and uncover new insights into the dynamics of the universe.

8.6 Ethical and Societal Implications

The discovery of dark energy raises important ethical and societal questions about the nature of scientific inquiry, the dissemination of knowledge, and the allocation of resources for scientific research.

8.6.1 Funding and Support for Research

Given the fundamental importance of dark energy to our understanding of the cosmos, it is essential to prioritize research funding and support for experiments and observatories dedicated to its study. Ensuring continued investment in dark energy research will require collaboration between governments, funding agencies, and the scientific community to address some of the most pressing questions in cosmology and theoretical physics.

8.6.2 Education and Public Engagement

Communicating the significance of dark energy research to the public is crucial for fostering interest and understanding in science and astronomy. By engaging with the public through outreach events, educational programs, and media outreach, scientists can inspire curiosity and appreciation for the wonders of the universe, helping to build a more scientifically literate society.

8.7 Conclusion

Dark energy stands as one of the most profound and enigmatic phenomena in the universe, shaping the fate of the cosmos and challenging our understanding of fundamental physics. From its discovery to its profound implications for the future of the universe, dark energy continues to captivate the imagination of scientists and the public alike.

As we continue to probe the mysteries of the cosmos, the quest to understand dark energy remains one of the greatest scientific endeavors of our time. Through interdisciplinary collaboration, technological innovation, and a relentless spirit of inquiry, scientists are poised to unlock the secrets of dark energy and illuminate the hidden depths of the universe. In doing so, we may uncover new insights into the nature of space, time, and the cosmos, reshaping our understanding of the universe and our place within it for generations to come.

Chapter 9: The Structure of the Universe: Clusters, Superclusters, and Filaments

In the vast expanse of the cosmos, galaxies are not randomly distributed but instead organized into vast cosmic structures that span billions of light-years. From the largest galaxy clusters to the intricate cosmic web of filaments, the structure of the universe offers profound insights into its evolution and the underlying laws of physics. In this chapter, we embark on a journey to explore the architecture of the cosmos, from the cosmic web to the clusters and superclusters that populate the universe.

9.1 The Cosmic Web: An Intricate Tapestry of Filaments

At the largest scales, the universe is arranged into a vast cosmic web—a complex network of filaments, voids, and clusters that spans the entirety of space. The cosmic web is the result of the gravitational collapse of primordial fluctuations in the density of matter, giving rise to a rich structure that astronomers continue to study and unravel.

9.1.1 Theoretical Foundations: From Inflation to Structure Formation

The formation of the cosmic web is rooted in the early moments of the universe's history, shortly after the Big Bang. According to the theory of cosmic inflation, the universe underwent a brief period of exponential expansion, stretching quantum fluctuations in the fabric of spacetime to cosmic scales. These fluctuations served as the seeds for the formation of cosmic structures through the process of gravitational instability, leading to the formation of filaments, clusters, and voids on large scales.

9.1.2 Simulations and Observations

Numerical simulations play a crucial role in understanding the formation and evolution of the cosmic web. By modeling the gravitational interactions of billions of particles over cosmic timescales, scientists can simulate the growth of cosmic structures and compare the results with observations from telescopes and surveys. These simulations provide valuable insights into the distribution of galaxies, dark matter, and other cosmic constituents, shedding light on the processes driving cosmic evolution.

9.2 Galaxy Clusters: Cosmic Cities of a Billion Suns

At the heart of the cosmic web lie galaxy clusters—immense structures composed of hundreds or thousands of galaxies bound together by gravity. Galaxy clusters are the largest gravitationally bound structures in the universe, serving as cosmic laboratories for studying the properties of galaxies, dark matter, and the underlying structure of spacetime.

9.2.1 Formation and Evolution

Galaxy clusters form through the hierarchical merger of smaller structures, such as galaxy groups and filaments, over cosmic timescales. As galaxies and dark matter halos merge and accrete, they gravitationally attract surrounding matter, leading to the growth of the cluster. Over time, galaxy clusters continue to evolve through mergers, accretion, and interactions with their cosmic environment, shaping the observed properties of galaxies and dark matter within the cluster.

9.2.2 Observational Signatures

Galaxy clusters exhibit a variety of observational signatures that reveal their presence and properties. These include the gravitational lensing of background galaxies, the X-ray emission from hot gas within the cluster, and the distribution of galaxies within the cluster itself. By studying these signatures, astronomers can infer the mass, size, and dynamics of galaxy clusters, providing valuable constraints on theories of structure formation and cosmology.

9.3 Superclusters: Cosmic Cities of a Trillion Suns

Beyond individual galaxy clusters lie superclusters—vast agglomerations of galaxy clusters and groups that stretch across hundreds of millions of light-years. Superclusters are the largest coherent structures in the universe, serving as nodes in the cosmic web and hubs of cosmic evolution.

9.3.1 The Great Attractor

One of the most famous superclusters in the local universe is the Great Attractor—a dense concentration of galaxies and galaxy clusters located in the direction of the constellation Centaurus. The Great Attractor exerts a powerful gravitational pull on the Milky Way and other nearby galaxies, causing them to move towards it at high speeds. Despite its name, the nature of the Great Attractor remains a subject of ongoing research and debate among astronomers.

9.3.2 The Sloan Great Wall

Another prominent supercluster is the Sloan Great Wall—a vast filamentary structure that extends over 1.3 billion light-years and contains thousands of galaxies. Discovered in the early 2000s by astronomers using data from the Sloan Digital Sky Survey, the Sloan Great Wall challenges our understanding of cosmic structure formation and the limits of gravitational clustering in the universe.

9.4 Cosmic Voids: The Dark Deserts of Space

Interwoven between the filaments and clusters of the cosmic web are vast regions known as cosmic voids—expansive regions of space that contain relatively few galaxies and matter compared to the surrounding filaments. Despite their emptiness, cosmic voids play a crucial role in shaping the large-scale structure of the universe and the distribution of galaxies within it.

9.4.1 Formation and Evolution

Cosmic voids form through the gravitational expansion of the universe, which stretches and thins the matter density in certain regions of space. As galaxies and clusters are drawn towards the dense filaments of the cosmic web, voids emerge in the regions left behind. Over cosmic timescales, voids continue to grow and evolve, shaping the distribution of matter and galaxies around them.

9.4.2 Observational Signatures

Cosmic voids exhibit distinctive observational signatures that distinguish them from other cosmic structures. These include the underdensity of galaxies and matter within the void, the presence of large-scale cosmic flows around the void boundary, and the suppression of galaxy formation and evolution within void regions. By studying these signatures, astronomers can probe the nature of voids and their role in cosmic structure formation.

9.5 The Cosmic Web: A Tapestry of Cosmic Evolution

The structure of the universe, from the largest galaxy clusters to the intricate filaments and voids of the cosmic web, offers profound insights into the evolution of the cosmos and the underlying laws of physics. By studying the distribution, properties, and dynamics of cosmic structures, astronomers can unravel the mysteries of cosmic evolution and the origins of the universe itself.

9.5.1 Cosmic Evolution and Dark Energy

The formation and evolution of cosmic structures are intimately linked to the properties of dark energy and the accelerated expansion of the universe. Dark energy influences the growth of cosmic structures through its effects on the cosmic expansion rate and the dynamics of gravitational collapse, shaping the observed distribution of galaxies and matter on large scales.

9.5.2 Future Directions in Cosmology

As observational techniques and theoretical models continue to advance, astronomers are poised to unlock new insights into the structure and evolution of the universe. From upcoming surveys and telescopes to simulations and theoretical frameworks, the future of cosmology promises to reveal new discoveries and deepen our understanding of the cosmos, from its earliest moments to its ultimate fate.

9.6 Conclusion

The structure of the universe, from the vast cosmic web to the intricate filaments and clusters that populate it, offers a window into the cosmic tapestry of cosmic evolution. By studying the distribution, properties, and dynamics of cosmic structures, astronomers are unraveling the mysteries of the cosmos and shedding light on the fundamental nature of the universe itself. As we continue to explore the wonders of the cosmos, the structure of the universe remains one of the most captivating and profound subjects in astronomy and astrophysics, inspiring wonder, curiosity, and awe in all who contemplate its mysteries.

Chapter 10: Cosmic Microwave Background Radiation: Echoes of the Big Bang

The universe is a treasure trove of cosmic secrets, and among its most enigmatic phenomena is the cosmic microwave background radiation (CMB). This faint glow, permeating the cosmos, offers a glimpse into the universe's infancy, serving as a relic of the Big Bang—the explosive event that birthed the cosmos. In this chapter, we embark on a journey to explore the cosmic microwave background radiation, uncovering its discovery, significance, and the profound insights it provides into the universe's early history and evolution.

10.1 In the Beginning: The Big Bang Theory

The story of the cosmic microwave background radiation begins with the Big Bang theory, which proposes that the universe originated from a hot, dense state approximately 13.8 billion years ago. According to this paradigm-shifting theory, the universe underwent a rapid expansion, cooling as it expanded, and giving rise to the diverse structures we observe today. The Big Bang theory, initially proposed by Georges Lemaître and later developed by scientists like George Gamow, Ralph Alpher, and Robert Herman, revolutionized our understanding of the universe's origins and laid the foundation for the study of the CMB.

10.2 The Discovery of the Cosmic Microwave Background

The cosmic microwave background radiation was discovered serendipitously in 1965 by Arno Penzias and Robert Wilson at Bell Labs in New Jersey, USA. In their quest to eliminate background noise from their radio antenna, they stumbled upon a persistent signal coming from all directions of the sky—a signal that was later identified as the CMB. Their discovery confirmed a key prediction of the Big Bang theory and earned them the Nobel Prize in Physics in 1978.

10.3 The Significance of the Cosmic Microwave Background

The discovery of the CMB was a watershed moment in cosmology, providing compelling evidence in support of the Big Bang theory and revolutionizing our understanding of the universe's origins. The significance of the CMB lies in its remarkable properties and the insights it offers into the universe's early history and evolution.

10.3.1 A Relic of the Big Bang

The cosmic microwave background radiation is a relic of the early universe, dating back to a time when the cosmos was just 380,000 years old—a mere fraction of its current age. At this epoch, the universe had cooled sufficiently for protons and electrons to combine and form neutral hydrogen atoms, allowing photons to travel freely through space. The CMB represents the "afterglow" of this primordial era, preserving a snapshot of the universe's temperature and density fluctuations at that time.

10.3.2 Probing the Universe's Composition

By studying the properties of the cosmic microwave background radiation, astronomers have been able to infer valuable information about the universe's composition and structure. The uniformity and isotropy of the CMB provide evidence for the cosmological principle—that the universe is homogeneous and isotropic on large scales. Moreover, the small-scale temperature fluctuations in the CMB encode information about the distribution of matter and energy in the early universe, including the presence of dark matter and dark energy.

10.4 Observations and Measurements of the Cosmic Microwave Background

Since its discovery, astronomers have made tremendous strides in observing and measuring the properties of the cosmic microwave background radiation, using a variety of ground-based, balloon-borne, and space-based instruments. These observations have provided valuable insights into the nature of the early universe and the processes that shaped its evolution.

10.4.1 The COBE Mission

One of the landmark missions in CMB research was the Cosmic Background Explorer (COBE) mission, launched by NASA in 1989. COBE's observations of the CMB's spectrum and anisotropies provided the first high-resolution map of the CMB and confirmed its blackbody nature, in excellent agreement with theoretical predictions. The mission's groundbreaking discoveries earned its principal investigators, John Mather and George Smoot, the Nobel Prize in Physics in 2006.

10.4.2 The WMAP Mission

Building upon the legacy of COBE, the Wilkinson Microwave Anisotropy Probe (WMAP), launched in 2001, further refined our understanding of the cosmic microwave background radiation. WMAP's observations of the CMB's temperature fluctuations with unprecedented precision allowed astronomers to measure key cosmological parameters, such as the age, geometry, and composition of the universe, with unprecedented accuracy. The mission provided compelling evidence in support of the standard cosmological model, known as the Lambda Cold Dark Matter (ΛCDM) model.

10.4.3 The Planck Satellite

The European Space Agency's Planck satellite, launched in 2009, represented the pinnacle of CMB research, providing the most detailed map of the CMB to date. Planck's observations of the CMB's temperature and polarization fluctuations have revolutionized our understanding of the universe's early history and evolution, constraining cosmological models and shedding light on the nature of dark matter, dark energy, and the cosmic inflation.

10.5 Polarization: Unveiling the Universe's Secrets

In addition to temperature fluctuations, the cosmic microwave background radiation also exhibits polarization—an intrinsic property of light that carries valuable information about the universe's early history and evolution. By studying the polarization patterns of the CMB, astronomers can probe the conditions of the early universe and test theories of cosmic inflation and the physics of the primordial plasma.

10.5.1 E-mode and B-mode Polarization

The polarization of the CMB can be decomposed into two components: E-mode polarization, which arises from density fluctuations in the primordial plasma, and B-mode polarization, which arises from primordial gravitational waves generated during cosmic inflation. Detecting B-mode polarization would provide smoking-gun evidence for inflation and offer valuable insights into the physics of the early universe.

10.5.2 Future Prospects

Future experiments and missions, such as the Atacama Cosmology Telescope (ACT), the Simons Observatory, and the Cosmic Origins Explorer (CORE), aim to further refine our understanding of the CMB and its implications for cosmology. By mapping the CMB's polarization with unprecedented sensitivity and resolution, these experiments promise to unlock new insights into the origins and evolution of the universe.

10.6 Conclusion

The cosmic microwave background radiation stands as a testament to the universe's remarkable journey from its fiery origins to the vast cosmos we observe today. From its serendipitous discovery to its role as a cornerstone of modern cosmology, the CMB continues to captivate the imagination of scientists and enthusiasts alike, offering a window into the universe's early history and evolution. As we peer back in time through the lens of the cosmic microwave background radiation, we are reminded of the incredible journey that has brought us to our current understanding of the cosmos.

10.6.1 Cosmological Implications

The cosmic microwave background radiation has profound cosmological implications, shaping our understanding of the universe's origins, composition, and fate. By studying its properties, astronomers have been able to test and refine our most fundamental theories of cosmology, including the Big Bang model, cosmic inflation, and the nature of dark matter and dark energy. The exquisite precision of CMB observations has allowed scientists to measure key cosmological parameters with unprecedented accuracy, providing valuable constraints on theoretical models of the universe.

10.6.2 The Legacy of Discovery

The discovery of the cosmic microwave background radiation represents a triumph of human curiosity and ingenuity, demonstrating the power of observation and experimentation in unraveling the mysteries of the cosmos. From its humble beginnings as a background noise in a radio antenna to its status as a cornerstone of modern cosmology, the CMB has inspired generations of scientists to probe the universe's deepest secrets and unlock its hidden truths.

10.6.3 Inspiring Future Generations

As we continue to explore the wonders of the cosmos, the cosmic microwave background radiation serves as a reminder of the boundless possibilities that await us in the pursuit of knowledge. Its discovery has inspired countless researchers to push the boundaries of our understanding, fueling a sense of wonder and curiosity about the universe and our place within it. By sharing the story of the CMB and its significance, we can inspire future generations to continue the quest for discovery and exploration, ensuring that the legacy of scientific inquiry endures for millennia to come.

In conclusion, the cosmic microwave background radiation stands as a testament to the universe's remarkable history and the ingenuity of humanity in unraveling its mysteries. From its discovery to its profound cosmological implications, the CMB continues to shape our understanding of the cosmos and inspire awe and wonder in all who contemplate its significance. As we journey forward into the depths of space and time, the echoes of the Big Bang remind us of the incredible journey that has brought us to this moment and the endless possibilities that lie ahead in our quest to explore the wonders of the universe.

Chapter 11: Galaxy Formation and Evolution

Galaxies are the building blocks of the universe, vast cosmic islands containing billions or even trillions of stars, as well as gas, dust, and dark matter. Understanding how these magnificent structures form and evolve is one of the central pursuits of modern astrophysics. In this chapter, we delve into the intricacies of galaxy formation and evolution, exploring the processes that shape the cosmos on the grandest scales.

11.1 The Early Universe: From Primordial Fluctuations to Proto-Galaxies

The seeds of galaxy formation were sown in the early universe, just moments after the Big Bang. Tiny fluctuations in the density of matter left over from the cosmic inflation epoch grew over time through gravitational instability, eventually giving rise to the first structures—proto-galaxies—approximately 100 million years after the Big Bang. These primordial structures laid the foundation for the rich tapestry of galaxies that would populate the cosmos in the eons to come.

11.1.1 Theoretical Frameworks: Hierarchical Structure Formation

Our understanding of galaxy formation is grounded in the framework of hierarchical structure formation, which posits that galaxies form through the merger and accretion of smaller structures, such as gas clouds and dwarf galaxies, over cosmic timescales. Gravity plays a central role in this process, drawing matter together to form larger and more massive structures, from small galaxy groups to massive galaxy clusters.

11.1.2 Observational Evidence: Probing the Cosmic Dawn

Observations of the distant universe provide valuable insights into the earliest stages of galaxy formation and evolution. Astronomers use powerful telescopes, such as the Hubble Space Telescope and the Atacama Large Millimeter/submillimeter Array (ALMA), to peer back in time and observe galaxies as they were in the distant past. By studying the light emitted by these ancient galaxies, astronomers can infer their properties, such as their size, shape, and chemical composition, and trace the evolution of galaxies over billions of years.

11.2 The Milky Way and Its Siblings: A Journey Through Cosmic Time

Our own galaxy, the Milky Way, offers a unique window into the processes of galaxy formation and evolution. Situated in the outskirts of the Local Group—a small cluster of galaxies that includes the Andromeda Galaxy and dozens of smaller satellite galaxies—the Milky Way provides astronomers with a laboratory for studying the formation and evolution of galaxies in a variety of environments.

11.2.1 The Formation of the Milky Way

The Milky Way likely formed through the hierarchical merger of smaller protogalactic fragments in the early universe. Over billions of years, these fragments coalesced to form the familiar spiral structure we see today, with a central bulge, a rotating disk of stars and gas, and a surrounding halo of old stars and globular clusters. Recent observations suggest that the Milky Way may have experienced a series of major merger events with smaller satellite galaxies, reshaping its structure and morphology.

11.2.2 The Milky Way's Siblings: Satellite Galaxies and Stellar Streams

The Milky Way is surrounded by a retinue of satellite galaxies—small, faint companions that orbit around the larger galaxy. These satellite galaxies provide valuable clues about the formation and evolution of the Milky Way and its larger counterparts. In addition to satellite galaxies, the Milky Way is also home to numerous stellar streams—elongated structures composed of stars that were torn from dwarf galaxies during gravitational interactions.

11.3 Galaxy Types and Morphology: From Spirals to Ellipticals

Galaxies come in a wide variety of shapes and sizes, ranging from majestic spiral galaxies with graceful arms to elliptical galaxies with smooth, featureless profiles. Understanding the diversity of galaxy types and their underlying morphology is essential for unraveling the processes that govern galaxy formation and evolution.

11.3.1 Spiral Galaxies: Cosmic Pinwheels

Spiral galaxies are characterized by their distinct spiral arms, which wind outward from a central bulge. These arms are sites of active star formation, where dense clouds of gas and dust coalesce to form new stars. Spiral galaxies, such as the Andromeda Galaxy and the Whirlpool

Galaxy, are among the most common galaxy types in the universe and offer valuable insights into the mechanisms driving galaxy evolution.

11.3.2 Elliptical Galaxies: Cosmic Spheroids

Elliptical galaxies, by contrast, lack the distinct spiral structure of their counterparts and instead exhibit a smooth, ellipsoidal profile. These galaxies are typically composed of older stars and contain little to no ongoing star formation. Elliptical galaxies come in a range of sizes and luminosities, from giant supergiants to compact dwarf ellipticals, and are thought to form through the merger and accretion of smaller galaxies over cosmic timescales.

11.4 Active Galactic Nuclei: Cosmic Powerhouses

At the hearts of many galaxies lies a supermassive black hole—a gravitational behemoth millions or even billions of times more massive than the Sun. When matter accretes onto these black holes, it can release vast amounts of energy in the form of radiation, creating what are known as active galactic nuclei (AGN). AGN come in a variety of forms, including quasars, blazars, and Seyfert galaxies, and are among the most luminous objects in the universe.

11.4.1 Quasars: Cosmic Beacons

Quasars are the most luminous objects in the universe, emitting vast amounts of energy across the electromagnetic spectrum. These enigmatic objects are thought to be powered by supermassive black holes at the centers of distant galaxies, where intense gravitational forces drive the infall of matter into an accretion disk surrounding the black hole. As the matter spirals inward, it heats up to temperatures of millions of degrees, releasing copious amounts of radiation in the process.

11.4.2 Blazars: Cosmic Jets

Blazars are a type of AGN characterized by powerful jets of relativistic particles that emanate from the vicinity of the supermassive black hole. These jets, which can extend for thousands of light-years, emit radiation across the entire electromagnetic spectrum, from radio waves to gamma rays. Blazars are among the most energetic objects in the universe and are thought to play a key role in shaping the evolution of galaxies and their surrounding environments.

11.5 Galaxy Mergers and Interactions: Cosmic Collisions

Galaxy mergers and interactions are common occurrences in the universe, driving the evolution of galaxies and shaping their morphology and structure. When galaxies collide, gravitational forces distort their shapes and trigger bursts of star formation, leading to the creation of new stellar populations and the redistribution of gas and dust.

11.5.1 The Antennae Galaxies: A Cosmic Tango

The Antennae Galaxies, located approximately 45 million light-years away in the constellation Corvus, provide a stunning example of galaxy interaction in action. These two spiral galaxies are in the process of merging, their spiral arms stretching out like antennae as they gravitational dance with one another. The collision between these galaxies has triggered intense star formation, leading to the formation of thousands of young, blue stars in the region where the two galaxies overlap.

11.5.2 The Mice Galaxies: A Cosmic Collision

The Mice Galaxies, also known as NGC 4676, are a pair of interacting galaxies located approximately 300 million light-years away in the constellation Coma Berenices. These galaxies are undergoing a dramatic collision that has stretched their spiral arms and triggered a burst of star formation. The gravitational forces generated by the interaction have caused gas and dust to be stripped from the galaxies, creating long tidal tails that extend for tens of thousands of light-years into space.

11.6 Galaxy Clusters: Cosmic Cities

Galaxy clusters are the largest gravitationally bound structures in the universe, containing hundreds to thousands of galaxies held together by their mutual gravitational attraction. These cosmic cities provide valuable insights into the formation and evolution of large-scale cosmic structures and the distribution of dark matter.

11.6.1 Abell 1689: A Cosmic Kaleidoscope

Abell 1689 is a massive galaxy cluster located approximately 2.2 billion light-years away in the constellation Virgo. This sprawling cluster contains thousands of galaxies and is one of the most massive structures in the universe. Abell

1689 is known for its spectacular gravitational lensing effects, which occur when the cluster's immense gravitational field bends and distorts the light from more distant objects behind it, creating arcs, rings, and multiple images of background galaxies.

11.6.2 The Bullet Cluster: A Cosmic Collision Course

The Bullet Cluster, also known as 1E 0657-56, is a galaxy cluster located approximately 3.8 billion light-years away in the constellation Carina. This cluster is named for the bullet-like shape of the X-ray emission observed in the cluster's hot gas, which is thought to result from a high-speed collision between two smaller galaxy clusters. The Bullet Cluster provides compelling evidence for the existence of dark matter, as the observed separation between the X-ray-emitting gas and the cluster's gravitational lensing signal indicates that dark matter is present and spatially separated from the ordinary matter.

11.7 The Role of Simulations: Modeling the Cosmos

Computer simulations play a crucial role in our understanding of galaxy formation and evolution, allowing astronomers to model the complex interplay of physical processes that govern the behavior of galaxies on cosmic scales. These simulations incorporate gravity, gas dynamics, star formation, and feedback processes, providing valuable insights into the mechanisms that drive galaxy evolution.

11.7.1 The Millennium Simulation: A Cosmic Virtual Universe

The Millennium Simulation is one of the most ambitious efforts to model the formation and evolution of the universe on large scales. This groundbreaking simulation, carried out by an international team of scientists, used supercomputers to track the evolution of over 10 billion dark matter particles within a cubic region of the universe spanning 2 billion light-years on a side. The simulation provided valuable insights into the distribution of galaxies, the formation of cosmic structures, and the growth of dark matter halos over cosmic time.

11.7.2 Illustris Simulation: Painting the Cosmic Landscape

The Illustris simulation is another groundbreaking effort to model the universe on large scales, incorporating not only dark matter but also gas, stars, and supermassive black holes. This state-of-the-art simulation, carried out by a team of researchers from around the world, traced the evolution of over 40,000 galaxies within a cubic region of the universe spanning 350 million light-years on a side. The simulation produced stunningly realistic images of cosmic structures, revealing the intricate web of filaments, clusters, and voids that make up the cosmic web.

11.8 Future Prospects: Unraveling the Mysteries of Galaxy Formation

As our understanding of galaxy formation and evolution continues to evolve, astronomers are poised to make even more groundbreaking discoveries in the years to come. Future observatories, such as the James Webb Space Telescope (JWST), the Square Kilometre Array (SKA), and the Large Synoptic Survey Telescope (LSST), promise to revolutionize our understanding of the cosmos, providing unprecedented sensitivity and resolution across the electromagnetic spectrum.

11.8.1 The James Webb Space Telescope: Peering Into the Cosmic Dawn

The James Webb Space Telescope (JWST), set to launch in the coming years, will be the premier observatory for studying the early universe and the formation of galaxies. With its large aperture and infrared capabilities, JWST will peer back in time to observe the first galaxies that formed in the aftermath of the Big Bang, shedding light on the processes that drove cosmic reionization and the transition from the cosmic dark ages to the luminous universe we see today.

11.8.2 The Square Kilometre Array: Mapping the Cosmic Web

The Square Kilometre Array (SKA), currently under construction in Australia and South Africa, will be the world's largest and most powerful radio telescope when completed. With its unprecedented sensitivity and resolution, SKA will revolutionize our understanding of galaxy formation and evolution, probing the cold gas and magnetic fields that govern the behavior of galaxies on cosmic scales. SKA will also study the distribution of neutral hydrogen gas in the early universe, providing valuable insights into the processes that drove the formation of the first galaxies.

11.8.3 The Large Synoptic Survey Telescope: A Cosmic Movie Camera

The Large Synoptic Survey Telescope (LSST), currently under construction in Chile, will conduct the most comprehensive survey of the night sky ever undertaken. With its wide field of view and rapid cadence, LSST will monitor billions of galaxies over the course of a decade, capturing transient events such as supernovae, gravitational lensing events, and tidal disruptions. LSST will provide valuable data on the distribution and evolution of galaxies, shedding light on the processes that govern their formation and evolution over cosmic time.

In conclusion, galaxy formation and evolution are among the most fascinating and complex phenomena in the universe. From the primordial fluctuations of the early universe to the majestic structures we observe today, galaxies offer a window into the cosmic past and the processes that have shaped the cosmos over billions of years. As our understanding of galaxy formation continues to evolve, we are poised to unlock even more secrets of the universe, unraveling the mysteries of cosmic evolution and our place within it.

Chapter 12: The Expanding Universe: Hubble's Law and Beyond

The concept of the expanding universe represents one of the most profound discoveries in modern cosmology, reshaping our understanding of the cosmos and our place within it. In this chapter, we embark on a journey to explore the evidence for cosmic expansion, the implications of Hubble's law, and the ongoing quest to unravel the mysteries of the expanding universe.

12.1 Prelude to Discovery: A Brief History of Cosmology

Before delving into the specifics of cosmic expansion, it's essential to understand the historical context that laid the groundwork for this revolutionary discovery. From ancient civilizations' cosmological myths and beliefs to the groundbreaking theories of Albert Einstein and Edwin Hubble, the story of cosmology is a tale of curiosity, inquiry, and discovery.

12.1.1 The Static Universe: Einstein's Blunder

In the early 20th century, the prevailing view among scientists was that the universe was static and unchanging—a cosmic island adrift in an infinite sea of space. This view was encapsulated in Albert Einstein's general theory of relativity, which suggested that the universe should either be expanding or contracting under the influence of gravity. To preserve a static universe, Einstein introduced the cosmological constant—an additional term in his equations intended to counteract the effects of gravity and maintain cosmic equilibrium.

12.1.2 The Birth of Modern Cosmology: Hubble's Revelation

The static universe hypothesis was upended in the 1920s when the astronomer Edwin Hubble made a series of groundbreaking observations that would forever change our understanding of the cosmos. Using the newly constructed 100-inch Hooker Telescope at Mount Wilson Observatory in California, Hubble observed distant galaxies and found that their spectral lines were shifted towards the red end of the spectrum—a phenomenon known as redshift. This discovery provided compelling evidence that the universe was not static but instead expanding—a revelation that would reshape our cosmic worldview.

12.2 Hubble's Law: The Key to Cosmic Expansion

Hubble's discovery of the redshift-distance relationship, now known as Hubble's law, provided astronomers with a powerful tool for measuring the expansion rate of the universe and probing its large-scale structure. The law states that the recessional velocity of a galaxy is proportional to its distance from Earth, with more distant galaxies receding at higher velocities—a direct consequence of cosmic expansion.

12.2.1 The Hubble Constant: Measuring the Rate of Expansion

At the heart of Hubble's law lies the Hubble constant, denoted by the symbol H_0, which represents the current rate of cosmic expansion. Determining the precise value of the Hubble constant has been a long-standing challenge for astronomers, requiring sophisticated observational techniques and theoretical models. Recent measurements from the Hubble Space Telescope, combined with observations of distant supernovae and the cosmic microwave background radiation, have converged on a value of approximately 70 kilometers per second per megaparsec—a rate that implies the universe is expanding at an accelerating pace.

12.2.2 The Age of the Universe: From Hubble's Constant to Cosmic History

The Hubble constant not only provides insight into the rate of cosmic expansion but also serves as a crucial parameter for estimating the age of the universe. By inversely scaling the Hubble constant to determine the time it would take for galaxies to recede from one another, astronomers can infer the age of the universe. Current estimates place the age of the universe at approximately 13.8 billion years—a figure consistent with independent measurements from studies of the cosmic microwave background radiation and the ages of the oldest stars and globular clusters.

12.3 The Big Bang Model: A Cosmic Genesis

The discovery of cosmic expansion provided compelling support for the Big Bang model—a cosmological framework that describes the universe's evolution from a hot, dense state to its current form. According to the Big Bang model, the universe began as a singularity—a point of infinite density and temperature—from which it rapidly expanded and cooled over billions of years.

12.3.1 Primordial Nucleosynthesis: Forging the Elements

In the earliest moments of the universe's history, known as the epoch of primordial nucleosynthesis, protons and neutrons combined to form the lightest elements—hydrogen, helium, and lithium. This process, which occurred approximately three minutes after the Big Bang, left a lasting imprint on the elemental composition of the universe, with roughly 75% of its mass consisting of hydrogen and 25% of helium, by mass.

12.3.2 The Cosmic Microwave Background Radiation: Echoes of the Big Bang

Another cornerstone of the Big Bang model is the cosmic microwave background radiation (CMB)—a faint glow that permeates the cosmos and serves as a relic of the universe's early history. The CMB was first observed in 1965 by Arno Penzias and Robert Wilson and has since been studied in great detail by a variety of ground-based, balloon-borne, and space-based observatories. The properties of the CMB, including its uniformity and isotropy, provide compelling evidence in support of the Big Bang model and offer valuable insights into the universe's early evolution.

12.4 Beyond Hubble's Law: Probing the Nature of Cosmic Expansion

While Hubble's law provides a straightforward relationship between a galaxy's distance and its recession velocity, the nature of cosmic expansion is far more complex. Recent observations have revealed that the rate of cosmic expansion is accelerating—a phenomenon attributed to dark energy, a mysterious form of energy that pervades the universe and counteracts the effects of gravity.

12.4.1 Dark Energy: The Cosmic Antigravity

Dark energy represents one of the most profound mysteries in modern cosmology, comprising roughly 70% of the total energy density of the universe. Its existence was first inferred from observations of distant supernovae, which revealed that the universe's expansion rate is increasing over time. The nature of dark energy remains poorly understood, with proposed explanations ranging from a cosmological constant to dynamic fields that evolve over cosmic time.

12.4.2 The Fate of the Universe: Expansion, Contraction, or Something Else?

The discovery of cosmic acceleration has profound implications for the fate of the universe. Depending on the nature of dark energy and other cosmological parameters, the universe may continue to expand indefinitely, eventually succumbing to a "big freeze" as galaxies drift farther and farther apart. Alternatively, the universe could experience a "big crunch," collapsing under the influence of gravity, or undergo a "big rip," tearing apart as dark energy's repulsive force overwhelms gravity's pull.

12.5 Future Prospects: Peering Deeper into the Cosmic Unknown

As our understanding of cosmic expansion continues to evolve, astronomers are poised to make even more groundbreaking discoveries in the years to come. Future observatories, such as the James Webb Space Telescope (JWST), the Euclid mission, and the Large Synoptic Survey Telescope (LSST), promise to revolutionize our understanding of dark energy, cosmic acceleration, and the fundamental nature of the universe.

12.5.1 The James Webb Space Telescope: Unveiling the Cosmic Dawn

The James Webb Space Telescope (JWST), scheduled to launch in the coming years, will be a powerful tool for studying the early universe and probing the nature of cosmic expansion. With its infrared capabilities, JWST will peer back in time to observe the first galaxies that formed in the aftermath of the Big Bang, shedding light on the processes that drove cosmic reionization and the transition from the cosmic dark ages to the luminous universe we see today. By studying the properties of galaxies in the early universe, JWST will provide valuable insights into the formation and evolution of cosmic structures and the interplay between dark matter, dark energy, and ordinary matter.

12.5.2 The Euclid Mission: Mapping the Dark Universe

The Euclid mission, led by the European Space Agency (ESA), aims to map the three-dimensional distribution of galaxies in the universe with unprecedented precision. By measuring the shapes and redshifts of galaxies across vast cosmic volumes, Euclid will create a detailed map of cosmic structure and probe the nature of dark energy and cosmic acceleration. The mission will also study the growth of large-scale structure in the universe, shedding light on the processes that govern galaxy formation and evolution.

12.5.3 The Large Synoptic Survey Telescope: A Cosmic Movie Camera

The Large Synoptic Survey Telescope (LSST), currently under construction in Chile, will conduct the most comprehensive survey of the night sky ever undertaken. With its wide field of view and rapid cadence, LSST will monitor billions of galaxies over the course of a decade, capturing transient events such as supernovae, gravitational lensing events, and tidal disruptions. LSST will provide valuable data on the distribution and evolution of galaxies, shedding light on the processes that govern their formation and evolution over cosmic time.

12.6 The Mysteries of Cosmic Expansion: Questions for the Future

Despite significant progress in our understanding of cosmic expansion, many mysteries remain unresolved. Key questions include the nature of dark energy, the fate of the universe, and the origin of cosmic acceleration. Addressing these questions will require further observational and theoretical advances, as well as new technologies and instruments capable of probing the cosmos on ever larger scales.

12.6.1 Dark Energy: A Fundamental Puzzle

One of the central mysteries of modern cosmology is the nature of dark energy—a mysterious force that permeates the universe and drives its accelerated expansion. Understanding the origin and properties of dark energy represents one of the greatest challenges facing astronomers today, requiring new observational techniques, theoretical models, and experimental tests.

12.6.2 The Fate of the Universe: Expansion, Contraction, or Something Else?

The ultimate fate of the universe remains uncertain, hinging on the delicate balance between the attractive force of gravity and the repulsive force of dark energy. Depending on the values of key cosmological parameters, the universe may continue to expand indefinitely, eventually succumbing to a "big freeze" as galaxies drift farther and farther apart. Alternatively, the universe could experience a "big crunch," collapsing under the influence of gravity, or undergo a "big rip," tearing apart as dark energy's repulsive force overwhelms gravity's pull.

12.6.3 Cosmic Acceleration: A Window into the Unknown

The discovery of cosmic acceleration has opened up new avenues of research into the fundamental nature of the universe. By studying the effects of dark energy on the large-scale structure of the cosmos, astronomers hope to gain insight into the underlying physics driving cosmic expansion and uncover new laws of nature that govern the universe on the grandest scales.

12.7 In Conclusion: The Expanding Universe and Our Cosmic Journey

The concept of the expanding universe represents one of the most profound insights in the history of science, reshaping our understanding of the cosmos and our place within it. From the pioneering observations of Edwin Hubble to the recent discoveries of cosmic acceleration, the story of cosmic expansion is a testament to human curiosity, ingenuity, and perseverance. As we continue to explore the mysteries of the expanding universe, we are poised to unlock even more secrets of the cosmos, revealing the fundamental laws that govern its evolution and our cosmic journey through space and time.

Chapter 13: Multiverse Theory: Exploring Other Realms

The concept of the multiverse represents one of the most mind-bending ideas in modern cosmology—a notion that suggests our universe may be just one of many, each with its own laws of physics, constants, and dimensions. In this chapter, we embark on a journey to explore the tantalizing possibility of the multiverse, delving into the theoretical frameworks, observational evidence, and philosophical implications of this radical idea.

13.1 Theoretical Foundations: From Inflation to Parallel Universes

The concept of the multiverse emerges from the fertile ground of theoretical physics, where ideas such as inflationary cosmology and string theory provide the scaffolding for exploring the possibility of other universes beyond our own.

13.1.1 Inflationary Cosmology: Seeds of the Multiverse

Inflationary cosmology, proposed by physicist Alan Guth in the 1980s, suggests that the early universe underwent a rapid period of exponential expansion just moments after the Big Bang. This inflationary phase, driven by a hypothetical scalar field, would have stretched the fabric of spacetime, smoothing out irregularities and setting the stage for the formation of galaxies and large-scale structure. While inflation provides a compelling explanation for the observed uniformity and flatness of the universe, it also predicts the existence of regions where inflation continues indefinitely, giving rise to an infinite multiverse.

13.1.2 String Theory: The Landscape of Possibilities

String theory, a theoretical framework that seeks to unify the fundamental forces of nature, offers another avenue for exploring the multiverse.

According to string theory, the universe is composed of tiny, vibrating strings of energy, whose different vibrational modes give rise to the particles and forces we observe. String theory also predicts the existence of additional spatial dimensions beyond the familiar three dimensions of space and one dimension of time, opening up the possibility of a vast landscape of possible universes with different dimensionalities and physical properties.

13.2 Types of Multiverse: Classification and Characteristics

The multiverse comes in various flavors, each with its own set of properties and implications for our understanding of reality. From the cosmological multiverse to the quantum multiverse, these different types of multiverses offer distinct avenues for exploring the nature of existence.

13.2.1 The Cosmological Multiverse: Bubble Universes and Eternal Inflation

In the context of inflationary cosmology, the cosmological multiverse consists of an infinite number of "pocket" or "bubble" universes, each with its own set of physical laws and constants. These bubble universes are thought to form as regions of spacetime undergo inflationary expansion, separating from one another and evolving independently over cosmic time. While each bubble universe may have different properties, they all share a common origin in the same underlying inflationary field.

13.2.2 The Many-Worlds Interpretation: Quantum Branching and Parallel Realities

In the realm of quantum mechanics, the Many-Worlds Interpretation (MWI) suggests that every quantum event results in the creation of multiple parallel universes, each corresponding to a different outcome of the event. According to MWI, these parallel universes coexist alongside our own, branching off from each other with every quantum interaction. While MWI remains a controversial interpretation of quantum mechanics, it offers a striking vision of reality in which every possible outcome of every quantum measurement plays out in a separate universe.

13.3 Observational Evidence: Probing the Multiverse Hypothesis

While the concept of the multiverse may seem like pure speculation, some scientists argue that there are observational signatures that could provide indirect evidence for its existence. From anomalies in the cosmic microwave background radiation to the properties of the fundamental constants of nature, these potential clues offer tantalizing hints of other universes beyond our own.

13.3.1 Cosmic Microwave Background: Imprints of Other Universes?

One potential avenue for probing the multiverse hypothesis lies in the cosmic microwave background radiation (CMB)—the faint glow left over from the Big Bang. Tiny variations in the temperature and polarization of the CMB could provide clues about the existence of other universes beyond our own. For example, anomalies in the statistical properties of the CMB could be interpreted as evidence of collisions between our universe and neighboring bubble universes, leaving behind imprints in the cosmic microwave background.

13.3.2 The Anthropic Principle: Fine-Tuning and Cosmic Coincidences

The anthropic principle, which suggests that the observed properties of the universe must be compatible with the existence of observers, offers another perspective on the multiverse hypothesis. According to the anthropic principle, the fundamental constants of nature and the laws of physics may be finely tuned to allow for the emergence of life—an idea known as the "fine-tuning problem." Proponents of the multiverse argue that the existence of multiple universes with different properties could provide a natural explanation for this apparent fine-tuning, as observers would necessarily find themselves in a universe conducive to their existence.

13.4 Philosophical Implications: The Nature of Reality and the Meaning of Existence

The concept of the multiverse raises profound questions about the nature of reality and our place within it. From the nature of consciousness to the existence of free will, the multiverse hypothesis challenges many deeply held assumptions about the universe and our role in shaping it.

13.4.1 The Nature of Consciousness: Minds in Many Worlds

One of the most intriguing philosophical implications of the multiverse hypothesis is its potential impact on our understanding of consciousness. If the multiverse is real, then there may be countless versions of ourselves spread across different universes, each experiencing its own unique set of experiences and events. This idea raises questions about the nature of identity, self-awareness, and the nature of reality itself.

13.4.2 The Meaning of Existence: A Universe Without Purpose?

The concept of the multiverse also challenges traditional notions of cosmic purpose and meaning. If the multiverse is infinite, with an infinite number of universes, then the concept of a single, overarching purpose or meaning may lose its significance. Instead, the multiverse suggests a vast and diverse tapestry of existence, where every possibility is realized somewhere, leading to a universe without a predetermined endpoint or ultimate goal.

13.5 Future Prospects: Exploring the Multiverse Frontier

As our understanding of cosmology continues to evolve, scientists are actively pursuing avenues to test and explore the multiverse hypothesis. From experimental probes to theoretical investigations, these efforts promise to shed light on one of the most profound questions in modern science—whether our universe is but one among many.

13.5.1 Experimental Probes: Searching for Multiverse Signatures

Despite the theoretical appeal of the multiverse hypothesis, experimental evidence remains elusive. However, scientists are actively searching for observational signatures that could provide indirect evidence for the existence of other universes. From high-energy particle colliders to precision measurements of cosmic phenomena, these experimental probes offer hope for uncovering the secrets of the multiverse.

13.5.2 Collider Experiments: Probing the Nature of Reality

One approach to testing the multiverse hypothesis involves high-energy particle colliders, such as the Large Hadron Collider (LHC) at CERN. By smashing particles together at near-light speeds, these experiments can recreate conditions similar to those that existed in the early universe, allowing scientists to study the fundamental forces and particles that govern our reality. While collider experiments may not directly detect other universes, they could reveal clues about the underlying structure of spacetime and the nature of extra dimensions predicted by string theory.

13.5.3 Cosmic Observations: Hunting for Multiverse Clues

Another avenue for testing the multiverse hypothesis lies in observations of cosmic phenomena, such as the cosmic microwave background radiation (CMB) and the distribution of galaxies. By searching for anomalies or unexpected patterns in these cosmic observables, astronomers hope to uncover hints of other universes beyond our own. For example, deviations from the expected statistical properties of the CMB could suggest interactions between our universe and neighboring bubble universes, leaving behind detectable signatures in the cosmic microwave background.

13.5.4 Quantum Experiments: Probing the Foundations of Reality

Quantum experiments offer another promising avenue for exploring the multiverse hypothesis. By studying the behavior of particles at the quantum level, scientists hope to uncover clues about the underlying structure of spacetime and the nature of reality itself. For example, experiments involving quantum entanglement—the phenomenon where particles become correlated in such a way that the state of one particle is instantly correlated with the state of another, regardless of the distance between them—could provide insights into the interconnectedness of different regions of the multiverse.

13.6 The Multiverse Debate: Controversies and Challenges

While the multiverse hypothesis has captured the imagination of scientists and the public alike, it remains a subject of considerable debate and controversy within the scientific community. Critics argue that the multiverse is inherently untestable and therefore falls outside the realm of empirical science. Others raise concerns about the philosophical implications of the multiverse, questioning whether it represents a legitimate scientific theory or merely a speculative idea.

13.6.1 Testability: Can the Multiverse Be Proven?

One of the primary criticisms of the multiverse hypothesis is its perceived lack of testability. Since other universes, by definition, lie beyond the observable horizon of our own, it may be impossible to directly detect or observe them. Critics argue that without empirical evidence, the multiverse remains a speculative idea rather than a scientifically viable theory. However, proponents counter that indirect observational signatures, combined with theoretical consistency and predictive power, could provide compelling support for the multiverse hypothesis.

13.6.2 Occam's Razor: Is the Multiverse Parsimonious?

Another criticism of the multiverse hypothesis concerns its adherence to the principle of Occam's Razor, which states that the simplest explanation that fits the data is usually the best. Critics argue that invoking an infinite number of universes to explain observed phenomena may violate Occam's Razor, as it introduces unnecessary complexity and undermines the principle of parsimony. However, proponents counter that the multiverse hypothesis may actually be the simplest explanation for certain phenomena, such as the apparent fine-tuning of the fundamental constants of nature, which would otherwise require ad hoc explanations.

13.7 Philosophical Reflections: The Nature of Reality and the Limits of Knowledge

Beyond the scientific debates and controversies, the concept of the multiverse raises profound questions about the nature of reality, the limits of human knowledge, and the relationship between science and philosophy. From the nature of existence to the possibility of ultimate truth, the multiverse hypothesis challenges us to rethink our assumptions about the universe and our place within it.

13.7.1 Reality and Perception: Do Other Universes Exist?

One of the central questions raised by the multiverse hypothesis is whether other universes exist independently of our own perception and observation. While the concept of the multiverse may seem far-fetched, some philosophers argue that it is no more implausible than other scientific theories, such as quantum mechanics or general relativity, which also challenge our intuitive understanding of reality. According to this view, the existence of other universes may be a natural consequence of the underlying laws of physics, even if we can never directly observe them.

13.7.2 The Limits of Knowledge: Can We Ever Know the Truth?

The multiverse hypothesis also raises questions about the limits of human knowledge and our ability to discern ultimate truths about the universe. If other universes exist beyond our own, then our understanding of reality may always be limited by our observational capabilities and theoretical frameworks. This idea challenges the notion that science can provide definitive answers to the deepest questions of existence, suggesting instead that our understanding of the cosmos may always be incomplete and provisional.

13.8 Conclusion: Navigating the Multiverse Landscape

In conclusion, the concept of the multiverse represents a bold and provocative idea that pushes the boundaries of our understanding of the cosmos. While the multiverse hypothesis remains speculative and controversial, it offers a fascinating glimpse into the nature of reality and the limits of human knowledge. Whether the multiverse ultimately proves to be a fruitful avenue for scientific inquiry or a philosophical puzzle that defies resolution, its exploration promises to captivate the imagination and challenge the intellect for generations to come.

Chapter 14: Gravitational Waves: Ripples in Spacetime

Gravitational waves represent one of the most elusive and extraordinary predictions of Albert Einstein's general theory of relativity. In this chapter, we embark on a journey to explore the fascinating world of gravitational waves, from their theoretical origins to their detection and their revolutionary implications for our understanding of the universe.

14.1 Theoretical Foundations: Einstein's Bold Prediction

In 1915, Albert Einstein published his general theory of relativity—a revolutionary framework that described gravity as the curvature of spacetime caused by the presence of mass and energy. Building on the mathematical formalism of his theory, Einstein made a bold prediction: the existence of gravitational waves—ripples in the fabric of spacetime that propagate through the universe at the speed of light.

14.1.1 Einstein's Field Equations: The Geometry of Gravity

At the heart of Einstein's theory of general relativity are his field equations—a set of differential equations that describe how matter and energy curve spacetime, and how this curvature in turn affects the motion of particles and the propagation of light. These equations provide a precise mathematical description of the gravitational force, linking the geometry of spacetime to the distribution of matter and energy within it.

14.1.2 The Nature of Gravitational Waves: Distortions in Spacetime

According to Einstein's theory, gravitational waves are produced by accelerating masses—such as orbiting stars or merging black holes—which create ripples in the curvature of spacetime that propagate outward in all directions. Unlike electromagnetic waves, which require a medium (such as air or water) to travel through, gravitational waves can pass through empty space unimpeded, carrying information about the violent astrophysical events that produced them.

14.2 Detection Efforts: A Century of Quest

Despite their theoretical prediction by Einstein over a century ago, gravitational waves remained elusive and undetected for much of the 20th century. However, scientists persevered in their quest to detect these elusive ripples, developing increasingly sophisticated instruments and experimental techniques in the hopes of observing the faint signals from distant cosmic sources.

14.2.1 Weber's Bar Detectors: Pioneering Efforts

In the 1960s, physicist Joseph Weber pioneered the first experimental efforts to detect gravitational waves using large aluminum bars suspended in a vacuum. Weber's detectors were designed to resonate in response to passing gravitational waves, producing measurable vibrations that could be recorded by sensitive detectors. While Weber claimed to have detected gravitational wave signals, subsequent experiments failed to replicate his results, casting doubt on his findings.

14.2.2 Interferometric Detectors: Precision Instruments for the 21st Century

In the late 20th and early 21st centuries, a new generation of gravitational wave detectors emerged, based on the principles of interferometry—a technique that uses the interference of light waves to measure minute changes in distance. Leading the way were the Laser Interferometer Gravitational-Wave Observatory (LIGO) in the United States and the Virgo interferometer in Italy, which began operation in the early 2000s with the goal of directly detecting gravitational waves.

14.3 Landmark Detection: The Dawn of Gravitational Wave Astronomy

After decades of development and refinement, the monumental breakthrough came on September 14, 2015, when the LIGO observatory made the first direct detection of gravitational waves. The signal, known as GW150914, was produced by the merger of two black holes located over a billion light-years away—a cataclysmic event that sent ripples through the fabric of spacetime, which were detected by LIGO's sensitive instruments.

14.3.1 GW150914: A Black Hole Ballet

The detection of GW150914 provided the first direct evidence of binary black hole mergers and confirmed a key prediction of general relativity. The signal revealed the characteristic "chirp" pattern expected from the inspiral and merger of two compact objects, followed by a brief burst of gravitational waves as the merged black hole settled into its final state. The observed properties of the signal matched the predictions of numerical relativity simulations, providing further confirmation of Einstein's theory.

14.3.2 Subsequent Discoveries: A New Era of Astronomy

In the years following the landmark detection of GW150914, LIGO and Virgo have made numerous additional detections of gravitational waves, opening up a new era of gravitational wave astronomy. These detections have included mergers of neutron stars, collisions between black holes and neutron stars, and even the detection of gravitational waves from a kilonova—a cataclysmic explosion that produced both gravitational waves and electromagnetic radiation, observed across the entire spectrum of light.

14.4 Scientific Implications: Probing the Cosmos with Gravitational Waves

The detection of gravitational waves has revolutionized our understanding of the universe, offering a new window into some of the most violent and energetic phenomena in the cosmos. From the nature of black holes to the origin of heavy elements, gravitational waves provide valuable insights into the fundamental processes that shape our universe.

14.4.1 Testing General Relativity: Einstein's Legacy

One of the primary goals of gravitational wave astronomy is to test the predictions of general relativity under extreme conditions. By observing the behavior of gravitational waves as they propagate through the universe, scientists can probe the nature of spacetime itself, testing Einstein's theory in regimes where its predictions differ from those of alternative theories of gravity.

14.4.2 Probing Black Holes: From Theory to Reality

Gravitational wave observations have provided unprecedented insights into the properties of black holes, confirming their existence and characterizing their masses, spins, and mergers. These observations have revealed the existence of "intermediate-mass" black holes, with masses between those of stellar-mass and supermassive black holes, as well as "extreme" black holes with spins close to the maximum allowed by general relativity.

14.4.3 Exploring the Cosmos: Beyond Black Holes

In addition to black hole mergers, gravitational wave astronomy has the potential to detect a wide range of other astrophysical phenomena, including the mergers of neutron stars, the collisions between black holes and neutron stars, and the gravitational waves produced by supernova explosions and cosmic strings. By observing these events, scientists hope to gain a deeper understanding of the processes that govern the evolution of galaxies, the formation of heavy elements, and the structure of the universe on the largest scales.

14.5 Future Prospects: The Next Frontier of Gravitational Wave Astronomy

As gravitational wave astronomy enters its second decade, scientists are poised to make even more groundbreaking discoveries with the next generation of detectors and instruments. From space-based observatories to third-generation ground-based detectors, these advancements promise to expand our understanding of the universe and unlock new mysteries of the cosmos.

14.5.1 LISA: The Next Frontier

The Laser Interferometer Space Antenna (LISA), a space-based gravitational wave observatory currently under development by ESA in collaboration with NASA, represents a major leap forward in gravitational wave astronomy. Scheduled for launch in the 2030s, LISA will consist of three spacecraft flying in a triangular formation, separated by millions of kilometers. By measuring the tiny changes in distance between the spacecraft caused by passing gravitational waves, LISA will be able to detect a wide range of gravitational wave sources with unprecedented sensitivity and accuracy.

14.5.2 Third-Generation Ground-Based Detectors: Pushing the Limits

While LIGO and Virgo continue to make groundbreaking discoveries, plans are already underway for the next generation of ground-based gravitational wave detectors. These third-generation detectors, such as the proposed Einstein Telescope in Europe and the Cosmic Explorer in the United States, will be even more sensitive than their predecessors, capable of detecting gravitational waves from even more distant and faint sources. With their improved sensitivity, these detectors will open up new windows into the universe, allowing scientists to explore previously inaccessible regions of spacetime.

14.6 Societal Implications: From Technology to Inspiration

Beyond their scientific importance, gravitational waves have profound societal implications, from the development of new technologies to the inspiration of future generations of scientists and engineers. By pushing the boundaries of what is possible, gravitational wave research has the potential to drive innovation, stimulate economic growth, and inspire the next generation of explorers and innovators.

14.6.1 Technological Innovations: From Precision Measurement to Quantum Computing

The pursuit of gravitational wave detection has led to the development of cutting-edge technologies with applications far beyond astronomy. From ultra-precise laser interferometry to advanced signal processing algorithms, the technologies developed for gravitational wave detection have found uses in fields ranging from telecommunications and navigation to medical imaging and quantum computing. By pushing the limits of what is possible, gravitational wave research continues to drive technological innovation and stimulate economic growth.

14.6.2 Educational Outreach: Inspiring the Next Generation

The discovery of gravitational waves has captured the imagination of people around the world, inspiring curiosity and wonder about the universe we inhabit. Through educational outreach programs, public lectures, and multimedia presentations, scientists and educators are sharing the excitement of gravitational wave astronomy with students, teachers, and the general public. By engaging with audiences of all ages and backgrounds, these outreach efforts are helping to inspire the next generation of scientists, engineers, and explorers.

14.7 Ethical Considerations: Gravitational Waves and Society

As gravitational wave research continues to advance, it is important to consider the ethical implications of our scientific endeavors. From questions of access and equity to issues of privacy and security, the pursuit of gravitational wave detection raises complex ethical challenges that require careful consideration and thoughtful engagement.

14.7.1 Access and Equity: Ensuring Inclusivity in Science

As gravitational wave research becomes increasingly globalized, it is essential to ensure that all countries and communities have equitable access to the benefits of scientific discovery. This includes not only access to data and resources, but also opportunities for participation and collaboration in scientific research. By promoting diversity and inclusivity in science, we can harness the full potential of gravitational wave astronomy to address some of the most pressing challenges facing humanity.

14.7.2 Privacy and Security: Safeguarding Sensitive Information

The sensitive nature of gravitational wave data raises important questions about privacy and security in the digital age. As gravitational wave detectors become more sensitive and capable of detecting a wider range of phenomena, it is essential to develop robust protocols for data sharing, storage, and security. By implementing strong safeguards and ethical guidelines, we can ensure that gravitational wave research remains transparent, accountable, and socially responsible.

14.8 Conclusion: The Cosmic Symphony of Gravitational Waves

In conclusion, the discovery of gravitational waves represents a triumph of human ingenuity and perseverance, opening up a new window into the universe and revolutionizing our understanding of gravity, spacetime, and the cosmos. From the theoretical predictions of Einstein to the monumental detections by LIGO and Virgo, the journey to unlock the secrets of gravitational waves has been a testament to the power of collaboration, innovation, and exploration. As we continue to explore the universe with ever greater precision and sensitivity, gravitational waves promise to reveal even more about the hidden workings of the cosmos, illuminating the cosmic symphony of spacetime and inspiring wonder and awe in all who contemplate its mysteries.

Chapter 15: Time and Space: Einstein's Theory of Relativity

Albert Einstein's theory of relativity revolutionized our understanding of time, space, and gravity, reshaping the very fabric of the universe. In this chapter, we delve into the profound insights of Einstein's theory, exploring its implications for our perception of reality, its applications in modern astrophysics, and its enduring legacy in the quest to unravel the mysteries of the cosmos.

15.1 Special Theory of Relativity: The Unity of Space and Time

Einstein's journey into the realm of relativity began in 1905 with the publication of his groundbreaking paper on the special theory of relativity. In this theory, Einstein proposed that the laws of physics are the same for all observers, regardless of their relative motion. Central to the special theory of relativity is the concept of spacetime—a four-dimensional continuum in which space and time are intertwined, forming the stage on which the drama of the universe unfolds.

15.1.1 The Principle of Relativity: Invariance of the Laws of Physics

At the heart of the special theory of relativity is the principle of relativity, which states that the laws of physics are invariant under transformations between inertial reference frames. This means that the fundamental equations of physics remain the same for observers in different states of motion, leading to the concept of relativistic symmetry and the abandonment of absolute space and time.

15.1.2 Time Dilation and Length Contraction: Relativistic Effects

One of the most remarkable consequences of Einstein's theory is the phenomenon of time dilation, whereby time appears to pass more slowly for observers in motion relative to those at rest. This effect, predicted by the Lorentz transformation equations, has been confirmed in numerous experiments and has practical implications for technologies such as GPS satellites, which must account for relativistic corrections to ensure accurate positioning.

15.2 General Theory of Relativity: The Geometry of Gravity

Building on the insights of the special theory of relativity, Einstein developed his general theory of relativity in 1915—a revolutionary framework that describes gravity as the curvature of spacetime caused by the presence of mass and energy. In this theory, gravity is not a force in the traditional sense but rather a manifestation of the curvature of spacetime, which guides the motion of objects through the cosmos.

15.2.1 The Equivalence Principle: Gravity as Geometry

Central to the general theory of relativity is the equivalence principle, which states that the effects of gravity are indistinguishable from those of acceleration. This principle led Einstein to the revolutionary idea that gravity is not a force propagated through space but rather a consequence of the curvature of spacetime—a concept known as the "geometric interpretation" of gravity.

15.2.2 The Curvature of Spacetime: Warping the Fabric of the Cosmos

According to Einstein's theory, massive objects such as stars and planets warp the fabric of spacetime around them, creating regions of curved spacetime known as gravitational fields. In the presence of these gravitational fields, the paths of light and matter are bent, leading to phenomena such as gravitational lensing and the deflection of starlight observed during solar eclipses.

15.3 Experimental Tests: Confirming Einstein's Vision

Despite its profound implications, Einstein's theory of relativity remained a bold and untested hypothesis until experimental evidence confirmed its predictions. Over the past century, numerous experiments and observations have validated the predictions of both the special and general theories of relativity, providing compelling evidence for the correctness of Einstein's vision of the cosmos.

15.3.1 Eclipses and Gravitational Lensing: Observing the Curvature of Spacetime

One of the earliest confirmations of general relativity came during the solar eclipse of 1919, when observations of starlight grazing the sun's limb revealed a slight deflection consistent with the predictions of Einstein's theory. Subsequent observations of gravitational lensing—the bending of light by massive objects—have further confirmed the curvature of spacetime predicted by general relativity, providing strong support for Einstein's gravitational theory.

15.3.2 Time Dilation and Particle Accelerators: Testing Relativistic Effects

In addition to astronomical observations, experiments conducted in particle accelerators have provided direct evidence for the predictions of special relativity, including time dilation and length contraction. By accelerating particles to near-light speeds, scientists have observed the relativistic effects predicted by Einstein's theory, confirming the validity of relativistic kinematics and dynamics.

15.4 Cosmological Implications: Relativity and the Cosmos

The insights of Einstein's theory of relativity have profound implications for our understanding of the universe on the largest scales, from the structure of spacetime to the evolution of the cosmos as a whole. By incorporating relativity into cosmological models, scientists have gained new insights into the nature of space, time, and the origin of the universe itself.

15.4.1 Big Bang Cosmology: From Singularity to Inflation

In the framework of general relativity, the universe is described as a dynamic, evolving entity whose history can be traced back to a hot, dense state known as the Big Bang. According to the Big Bang model, the universe began as a singularity—a point of infinite density and temperature—and has been expanding and cooling ever since. Subsequent developments, such as cosmic inflation, have provided additional support for the Big Bang model and its predictions for the large-scale structure of the universe.

15.4.2 Dark Energy and the Fate of the Universe: Relativistic Cosmology

Recent observations of the accelerating expansion of the universe have led to the discovery of dark energy—a mysterious form of energy that permeates all of spacetime and drives the universe apart. In the framework of general relativity, dark energy is thought to arise from the vacuum energy density of empty space, leading to a repulsive gravitational effect that counteracts the attractive force of gravity. The existence of dark energy poses profound questions about the ultimate fate of the universe and the nature of spacetime on cosmic scales.

15.5 The Legacy of Relativity: From Black Holes to Time Travel

Einstein's theory of relativity has left an indelible mark on our understanding of the universe and continues to inspire scientific inquiry across a wide range of disciplines. From the study of black holes to the exploration of the early universe, relativity provides a powerful framework for understanding the fundamental laws that govern the cosmos. In this section, we explore some of the diverse applications and ongoing research inspired by Einstein's theory of relativity.

15.5.1 Black Holes: Gravity's Ultimate Test

Black holes represent one of the most extreme predictions of general relativity—a region of spacetime where gravity is so intense that nothing, not even light, can escape its grasp. By studying the behavior of matter and light in the vicinity of black holes, scientists can test the predictions of general relativity in the most extreme gravitational environments imaginable. Observations of black hole mergers, accretion disks, and jets provide valuable insights into the nature of gravity and spacetime near these cosmic behemoths.

15.5.2 Gravitational Waves: Listening to the Universe

The recent detection of gravitational waves by instruments such as LIGO and Virgo has opened up a new frontier in astronomy, allowing scientists to listen to the universe as never before. By studying the signals emitted by merging black holes, neutron stars, and other cosmic phenomena, researchers can test the predictions of general relativity and probe the nature of spacetime on the largest scales. Gravitational wave astronomy promises to revolutionize our understanding of the universe and uncover new mysteries hidden in the depths of spacetime.

15.5.3 Cosmological Models: Exploring the Universe's Origins

Relativity plays a central role in modern cosmology, providing the theoretical framework for understanding the origin, evolution, and fate of the universe. Cosmological models based on general relativity, such as the Lambda-CDM model, incorporate observations of the cosmic microwave background, galaxy clustering, and supernova distances to paint a detailed picture of the universe's history from the Big Bang to the present day. By refining these models and testing their predictions against new observations, scientists hope to unlock the secrets of the cosmos and answer fundamental questions about the nature of space, time, and reality itself.

15.5.4 Time Travel and Wormholes: Science Fiction or Reality?

One of the most intriguing aspects of relativity is its prediction of time dilation and the possibility of time travel under certain conditions. While practical time travel remains firmly in the realm of science fiction, the theoretical possibility of traversable wormholes—hypothetical shortcuts through spacetime that could connect distant regions of the universe—has captured the imagination of scientists and science fiction writers alike. While the existence of traversable wormholes remains speculative, their study provides valuable insights into the nature of spacetime and the fundamental laws of physics.

15.6 The Future of Relativity: New Frontiers in Physics

As we look to the future, relativity continues to inspire new avenues of research and exploration in physics. From the quest for a unified theory of gravity and quantum mechanics to the search for exotic phenomena such as gravitational waves and black hole mergers, relativity remains at the forefront of scientific inquiry. By pushing the boundaries of our understanding of space, time, and gravity, relativity promises to unlock new mysteries of the cosmos and illuminate the fundamental nature of the universe itself.

15.6.1 Quantum Gravity: Bridging the Divide

One of the greatest challenges facing modern physics is the reconciliation of general relativity with quantum mechanics—the two pillars of 20th-century physics that describe the behavior of the universe on vastly different scales. While general relativity provides a classical description of gravity on cosmic scales, quantum mechanics governs the behavior of particles on the subatomic level. The quest for a theory of quantum gravity that unifies these two frameworks remains one of the holy grails of theoretical physics, with potential implications for our understanding of the fundamental nature of reality.

15.6.2 Relativistic Astrophysics: Probing Extreme Environments

Relativity plays a crucial role in understanding the behavior of matter and energy in extreme astrophysical environments, from the intense gravitational fields of black holes to the high-energy processes occurring in supernova explosions and gamma-ray bursts. By combining observations from telescopes, particle detectors, and gravitational wave observatories, scientists can probe the most violent and energetic phenomena in the universe, testing the predictions of general relativity and uncovering new insights into the nature of spacetime and gravity.

15.6.3 Experimental Tests: Pushing the Limits of Relativity

Advances in experimental techniques and technologies continue to push the limits of relativity, allowing scientists to test its predictions with ever greater precision and accuracy. From laboratory experiments probing the fundamental constants of nature to astronomical observations of exotic phenomena in the distant universe, the quest to understand the nature of space, time, and gravity remains one of the most exciting and challenging endeavors in modern science. As we continue to explore the mysteries of the cosmos, relativity will undoubtedly play a central role in shaping our understanding of the universe and our place within it.

Chapter 16: Cosmic Inflation: Rapid Expansion in the Early Universe

In the quest to understand the origins and evolution of the universe, one of the most intriguing and revolutionary concepts to emerge is cosmic inflation. Proposed in the early 1980s, cosmic inflation suggests that the universe underwent an exponential expansion phase in the first fraction of a second after the Big Bang. In this chapter, we delve into the theory of cosmic inflation, its implications for our understanding of the cosmos, and the ongoing efforts to test its predictions.

16.1 Prelude to Inflation: The Big Bang Model

Before delving into cosmic inflation, it's crucial to understand the framework within which it operates: the Big Bang model. According to this model, the universe began as a hot, dense singularity approximately 13.8 billion years ago and has been expanding and cooling ever since. The Big Bang model successfully explains a wide range of observed phenomena, from the cosmic microwave background radiation to the abundance of light elements in the universe.

16.1.1 Problems with the Big Bang Model: The Horizon Problem and Flatness Problem

Despite its successes, the Big Bang model faces two major challenges known as the horizon problem and the flatness problem. The horizon problem arises from the fact that different regions of the observable universe have the same temperature despite never having been in causal contact with each other. The flatness problem relates to the fine-tuning required to explain the universe's nearly flat geometry observed today.

16.2 Enter Cosmic Inflation: Alan Guth's Vision

In the early 1980s, physicist Alan Guth proposed the theory of cosmic inflation as a solution to these problems. Guth's idea was that in the first moments after the Big Bang, the universe underwent a brief period of exponential expansion, stretching out microscopic quantum fluctuations to cosmic scales and homogenizing the distribution of matter and energy.

16.2.1 Guth's Epiphany: From Particle Physics to Cosmology

Guth's inspiration for cosmic inflation came from his background in particle physics and his study of phase transitions in the early universe. Drawing on ideas from grand unified theories (GUTs) and quantum field theory, Guth realized that a brief period of exponential expansion could solve many of the puzzles plaguing the Big Bang model.

16.2.2 Inflationary Potential: Energy Fields Driving Expansion

At the heart of Guth's inflationary model is the concept of an inflationary potential—a form of energy that drives the rapid expansion of the universe. According to the theory, this energy field is thought to be associated with a hypothetical scalar field known as the inflaton, which undergoes a phase transition as the universe cools, triggering the onset of inflation.

16.3 The Dynamics of Inflation: From Quantum Fluctuations to Cosmic Structures

During the inflationary epoch, quantum fluctuations in the inflaton field are stretched across the universe, generating tiny variations in the density of matter and energy. These fluctuations serve as the seeds for the formation of cosmic structures, such as galaxies and galaxy clusters, observed in the universe today. The rapid expansion of space also flattens the geometry of the universe, addressing the flatness problem of the Big Bang model.

16.3.1 Inflation and the Cosmic Microwave Background: Imprints of the Early Universe

One of the key predictions of cosmic inflation is the existence of faint patterns in the cosmic microwave background radiation—the relic radiation left over from the Big Bang. These patterns, known as primordial density fluctuations, are thought to arise from quantum fluctuations during the inflationary epoch and provide valuable clues about the early universe's conditions.

16.3.2 From Quantum to Cosmos: Testing Inflation's Predictions

Observational tests of cosmic inflation come from a variety of sources, including measurements of the cosmic microwave background, large-scale structure surveys, and studies of the polarization of light from the early universe. By comparing these observations to the predictions of inflationary models, scientists can determine whether inflation provides a viable explanation for the universe's observed properties.

16.4 Challenges and Extensions: Fine-Tuning and Multiverse Theories

Despite its successes, cosmic inflation also faces challenges and unanswered questions. One of the most significant challenges is the fine-tuning required to set up the initial conditions for inflation to occur. Additionally, some variants of inflationary models predict the existence of a multiverse—a vast ensemble of parallel universes with different properties—raising philosophical and theoretical questions about the nature of reality.

16.4.1 Eternal Inflation: Inflation Without End

In some inflationary models, the process of exponential expansion continues indefinitely in certain regions of spacetime, leading to the concept of eternal inflation. In an eternal inflationary scenario, new universes are constantly being created within a larger, inflating multiverse, each with its own distinct properties and evolution.

16.4.2 New Observational Frontiers: Probing the Multiverse

While the concept of a multiverse remains highly speculative, ongoing observational efforts may provide insights into its possible existence. By studying the cosmic microwave background, the distribution of galaxies, and other cosmological observables, scientists hope to uncover evidence for the presence of other universes beyond our own, shedding light on the ultimate origin and fate of the cosmos.

16.5 Conclusion: Inflation and the Cosmic Odyssey

In conclusion, cosmic inflation represents a bold and revolutionary idea that has transformed our understanding of the universe's early history.

From its humble beginnings as a solution to the puzzles of the Big Bang model to its potential implications for the nature of reality itself, inflation has captured the imagination of scientists and laypeople alike. As we continue to probe the mysteries of the cosmos, cosmic inflation promises to remain a central pillar of our cosmic odyssey, guiding us on our journey to unlock the secrets of the universe's origins and evolution.

Chapter 17: Quantum Cosmology: Bridging the Gap Between Micro and Macro Worlds

In the vast tapestry of the cosmos, the realms of quantum mechanics and cosmology converge, offering tantalizing glimpses into the fundamental nature of reality. Quantum cosmology seeks to unify the laws of quantum mechanics with the dynamics of the universe on cosmic scales, probing the mysteries of the early universe and the nature of spacetime itself. In this chapter, we embark on a journey into the realm of quantum cosmology, exploring its origins, theoretical frameworks, and implications for our understanding of the cosmos.

17.1 The Quantum Revolution: From Micro to Macro

The story of quantum cosmology begins with the quantum revolution of the early 20th century, when physicists such as Max Planck, Albert Einstein, and Niels Bohr revolutionized our understanding of the microscopic world. Quantum mechanics, the theory that describes the behavior of particles on the smallest scales, introduced a new set of rules and principles that challenged our classical notions of reality.

17.1.1 Planck's Quantum Hypothesis: A Quantum of Action

Max Planck's groundbreaking insight into the nature of blackbody radiation laid the foundation for quantum mechanics. In 1900, Planck proposed that energy is quantized, meaning it can only take on discrete values, rather than being continuous as classical physics assumed. This revolutionary idea paved the way for the development of quantum mechanics and transformed our understanding of the physical world.

17.1.2 Einstein's Photon: Light as Particles

Albert Einstein further advanced the quantum revolution with his theory of the photoelectric effect, in which he proposed that light consists of discrete packets of energy called photons. Einstein's work provided compelling evidence for the particle-like nature of light and helped establish the wave-particle duality—the idea that particles exhibit both wave-like and particle-like behavior—as a central tenet of quantum mechanics.

17.1.3 Bohr's Quantum Model: The Birth of Quantum Theory

Niels Bohr's pioneering model of the hydrogen atom, based on the quantization of electron orbits, represented a major milestone in the development of quantum theory. Bohr's model successfully explained the discrete spectral lines observed in atomic spectra and laid the groundwork for the modern understanding of atomic structure and the behavior of electrons in atoms.

17.2 The Birth of Quantum Cosmology: Wheeler's Vision

The emergence of quantum mechanics in the early 20th century paved the way for the development of quantum cosmology—a field dedicated to applying the principles of quantum mechanics to the study of the universe as a whole. The seeds of quantum cosmology were planted by physicists such as John Wheeler, who recognized the need for a quantum description of the early universe.

17.2.1 Wheeler's Quantum Foam: Spacetime as a Quantum Entity

John Wheeler's concept of "quantum foam" envisioned spacetime itself as a frothy, fluctuating sea of quantum fluctuations on the smallest scales. Wheeler's vision challenged the classical notion of spacetime as a smooth, continuous fabric and laid the foundation for quantum cosmological models that incorporate the quantum nature of spacetime.

17.2.2 The Wheeler-DeWitt Equation: Quantum Gravity in Action

Central to the development of quantum cosmology is the Wheeler-DeWitt equation—a theoretical framework that seeks to describe the quantum state of the universe as a whole. Formulated in the 1960s by Bryce DeWitt, the Wheeler-DeWitt equation represents an attempt to unify the principles of quantum mechanics with the dynamics of general relativity, providing a quantum description of the cosmos.

17.3 Quantum Fluctuations and the Early Universe: From Nothing to Something

Quantum cosmology suggests that the universe emerged from a state of primordial nothingness, where fluctuations in the quantum vacuum gave rise to the seeds of cosmic structure. These quantum fluctuations are thought to have played a crucial role in the formation of the universe's large-scale structure, seeding the formation of galaxies, galaxy clusters, and other cosmic structures observed today.

17.3.1 Inflationary Quantum Cosmology: Guth's Inflationary Universe

Building on the ideas of cosmic inflation, quantum cosmologists such as Alan Guth proposed that the rapid expansion of the universe during the inflationary epoch was driven by quantum fluctuations in the early universe. According to this model, quantum fluctuations in the inflaton field stretched out by cosmic inflation served as the seeds for the formation of cosmic structure, providing a quantum explanation for the universe's observed properties.

17.3.2 Quantum Origin of Cosmological Perturbations: Seeds of Structure

Quantum cosmology suggests that the density fluctuations observed in the cosmic microwave background radiation—the relic radiation left over from the Big Bang—arise from quantum fluctuations in the early universe. These primordial density fluctuations are thought to have been amplified by cosmic inflation, leading to the formation of the large-scale structure observed in the universe today.

17.4 Challenges and Controversies: The Quantum-Cosmological Frontier

Despite its theoretical elegance, quantum cosmology faces significant challenges and controversies. One of the major challenges is the formulation of a consistent quantum theory of gravity that can accurately describe the dynamics of the early universe. Additionally, the interpretation of quantum cosmological models and their implications for the nature of reality remain subjects of intense debate among physicists and cosmologists.

17.4.1 The Measurement Problem: Observing the Unobservable

One of the central challenges of quantum cosmology is the measurement problem—the question of how to reconcile the probabilistic nature of quantum mechanics with the deterministic evolution of the universe. In the context of quantum cosmology, this problem raises profound questions about the nature of observation, measurement, and reality itself, challenging our classical intuitions about the nature of the cosmos.

17.4.2 Multiverse Theories: A Quantum-Cosmological Landscape

Some quantum cosmological models predict the existence of a multiverse—a vast ensemble of parallel universes with different properties and evolutions. These multiverse theories arise from the quantum nature of the early universe and the dynamics of inflation, leading to the possibility of an infinite variety of universes coexisting within a larger quantum-cosmological landscape.

17.5 The Future of Quantum Cosmology: Unraveling the Mysteries of the Cosmos

As we look to the future, quantum cosmology promises to remain at the forefront of scientific inquiry, offering new insights into the fundamental nature of reality and the origins of the universe. By bridging the gap between the microscopic world of quantum mechanics and the macroscopic realm of cosmology, quantum cosmology holds the potential to unlock some of the deepest mysteries of the cosmos and reshape our understanding of the universe and our place within it.

Chapter 18: Future Directions in Cosmology: Challenges and Discoveries Ahead

As humanity's understanding of the cosmos continues to deepen, new horizons of exploration and discovery beckon. In this chapter, we embark on a journey into the future of cosmology, exploring the challenges, opportunities, and potential breakthroughs that lie ahead in our quest to unravel the mysteries of the universe.

18.1 The Quest for Dark Matter: Unraveling the Invisible Universe

One of the most pressing challenges in cosmology is the nature of dark matter—an elusive form of matter that does not emit, absorb, or reflect light and comprises approximately 27% of the universe's total mass-energy content. Despite decades of research, the identity of dark matter remains unknown. Future experiments, such as the Large Synoptic Survey Telescope (LSST) and the European Space Agency's Euclid mission, aim to map the distribution of dark matter on cosmic scales and shed light on its properties.

18.1.1 Particle Physics and Dark Matter: The Search for WIMPs

Many cosmologists believe that dark matter consists of weakly interacting massive particles (WIMPs), hypothetical particles that interact with ordinary matter only through gravity and the weak nuclear force. Experimental efforts to detect WIMPs are ongoing, with experiments such as the Cryogenic Dark Matter Search (CDMS) and the XENON Collaboration searching for signs of dark matter interactions in underground laboratories around the world.

18.1.2 Galactic Dynamics and Dark Matter: Probing the Invisible

Observations of the rotation curves of galaxies and the gravitational lensing of distant objects provide compelling evidence for the existence of dark matter. By studying the dynamics of galaxies and galaxy clusters, astronomers can infer the distribution of dark matter within them and constrain its properties. Future observations with next-generation telescopes and observatories promise to refine our understanding of dark matter and its role in shaping the cosmos.

18.2 Unveiling the Nature of Dark Energy: Understanding the Cosmic Acceleration

Another major mystery in cosmology is the nature of dark energy—a mysterious form of energy that permeates the universe and is driving its accelerated expansion. Dark energy accounts for approximately 68% of the universe's total energy density and poses profound challenges to our current understanding of fundamental physics. Future observations, such as those from the Dark Energy Survey (DES) and the

European Space Agency's Euclid mission, aim to map the distribution of dark energy and constrain its properties.

18.2.1 Cosmic Acceleration and the Fate of the Universe: A Cosmic Tug of War

The discovery of cosmic acceleration, based on observations of distant supernovae in the late 1990s, revolutionized our understanding of the universe's expansion. Dark energy is thought to be responsible for this accelerated expansion, counteracting the gravitational pull of matter and driving galaxies apart at an ever-increasing rate. Understanding the nature of dark energy is crucial for determining the ultimate fate of the universe—whether it will continue expanding indefinitely or eventually collapse in a "Big Crunch."

18.2.2 Modified Gravity Theories: Alternatives to Dark Energy

While dark energy remains the leading explanation for cosmic acceleration, alternative theories of gravity have been proposed to explain the observed phenomena without invoking dark energy. Modified gravity theories, such as f(R) gravity and MOND (Modified Newtonian Dynamics), modify Einstein's theory of general relativity on large scales and offer alternative explanations for the observed acceleration of the universe. Future observational tests and experiments will help distinguish between these competing theories and shed light on the true nature of cosmic acceleration.

18.3 Cosmic Dawn and the First Stars: Probing the Early Universe

One of the most exciting frontiers in cosmology is the study of cosmic dawn—the period in the universe's history when the first stars and galaxies began to form. By observing the cosmic microwave background radiation and distant galaxies, astronomers can probe the conditions of the early universe and unravel the mysteries of cosmic dawn. Future telescopes and observatories, such as the James Webb Space Telescope (JWST) and the Square Kilometre Array (SKA), promise to revolutionize our understanding of the early universe and the formation of the first cosmic structures.

18.3.1 The First Stars and Galaxies: Building Blocks of the Cosmos

The formation of the first stars and galaxies marked a crucial transition in the history of the universe, leading to the emergence of the cosmic structures we see today. Understanding the processes that drove the formation of these early cosmic objects is essential for unraveling the origins of galactic diversity and the evolution of cosmic structure over cosmic time. Future observations with next-generation telescopes will allow astronomers to peer deeper into the cosmic dawn and uncover the secrets of the universe's infancy.

18.3.2 Reionization and Cosmic Rebirth: Illuminating the Dark Ages

Another key epoch in cosmic history is the era of reionization—a period when the universe's neutral hydrogen gas was ionized by the intense radiation from the first stars and galaxies. Reionization transformed the universe from a dark, opaque state to a transparent, ionized state, paving the way for the formation of more complex structures such as galaxies and galaxy clusters. Future observations of distant galaxies and quasars will shed light on the process of reionization and its role in shaping the cosmic landscape.

18.4 Multimessenger Astronomy: Uniting Light, Matter, and Gravitational Waves

The emerging field of multimessenger astronomy promises to revolutionize our understanding of the universe by combining information from multiple sources, including electromagnetic radiation, cosmic rays, neutrinos, and gravitational waves. By detecting signals from cosmic phenomena using different messengers, astronomers can gain new insights into the most extreme and energetic events in the cosmos, such as supernova explosions, black hole mergers, and gamma-ray bursts. Future multimessenger observatories, such as the Cherenkov Telescope Array (CTA) and the Laser Interferometer Space Antenna (LISA), will open new windows onto the universe and uncover phenomena beyond the reach of traditional telescopes.

18.4.1 Gravitational Wave Astronomy: Listening to the Symphony of the Universe

The recent detection of gravitational waves by instruments such as LIGO and Virgo has ushered in a new era of gravitational wave astronomy, allowing scientists to listen to the universe as never before. Gravitational waves are ripples in spacetime produced by the most violent and energetic events in the cosmos, such as the mergers of black holes and neutron stars.

By studying these signals, astronomers can probe the nature of gravity, test the predictions of general relativity, and unlock the secrets of the universe's most extreme phenomena.

18.4.2 Neutrino Astronomy: Ghostly Messengers from the Depths of Space

Neutrinos are elusive, nearly massless particles that interact very weakly with matter, making them extremely difficult to detect. However, neutrinos play a crucial role in many astrophysical phenomena, including supernova explosions, gamma-ray bursts, and the formation of black holes. By detecting neutrinos from cosmic sources, astronomers can probe the most extreme environments in the universe and gain new insights into the processes driving cosmic evolution.

18.5 Conclusion: The Cosmic Odyssey Continues

As we look to the future, the field of cosmology stands on the brink of unprecedented discovery and exploration. From unraveling the mysteries of dark matter and dark energy to probing the cosmic dawn and the formation of the first stars and galaxies, cosmologists are poised to unlock the secrets of the universe on scales both large and small. With cutting-edge telescopes, observatories, and experimental facilities at their disposal, scientists are poised to push the boundaries of our cosmic knowledge and reshape our understanding of the cosmos.

18.5.1 The Role of Technology: Advancements in Observation and Experimentation

Advancements in technology have played a pivotal role in driving progress in cosmology. From the development of powerful telescopes and detectors to the construction of state-of-the-art laboratories and experimental facilities, technological innovations have enabled scientists to push the limits of observation and experimentation. Future breakthroughs in instrumentation and data analysis techniques promise to open new avenues for discovery and revolutionize our understanding of the universe.

18.5.2 Theoretical Innovations: Pushing the Boundaries of Understanding

In parallel with experimental advancements, theoretical innovations continue to push the boundaries of our understanding of the cosmos. The development of new mathematical models, computational techniques, and conceptual frameworks allows scientists to explore the universe's most fundamental questions, from the nature of dark matter and dark energy to the origin of cosmic structures and the ultimate fate of the universe. Interdisciplinary collaborations between theorists and experimentalists are driving progress in cosmology and fostering new insights into the nature of reality.

18.5.3 Education and Outreach: Inspiring the Next Generation of Explorers

As we embark on this cosmic odyssey, it's essential to inspire and educate the next generation of scientists and explorers. Through outreach programs, educational initiatives, and public engagement efforts, scientists can share the excitement of cosmological discovery with people of all ages and backgrounds, sparking curiosity and fostering a deeper appreciation for the wonders of the universe. By empowering future generations to pursue careers in science and exploration, we ensure that the quest to understand the cosmos will continue for generations to come.

Conclusion:

The Universe and Cosmology: Explore the Mysteries of the Cosmos, from the Big Bang Theory to the Latest Discoveries in Astronomy and Astrophysics, has taken us on a captivating journey through the depths of space and time, from the origins of the cosmos to the frontiers of modern cosmological research. Throughout our exploration, we have encountered a myriad of phenomena, theories, and discoveries that have reshaped our understanding of the universe and our place within it.

From the explosive birth of the universe in the Big Bang to the formation of galaxies, stars, and planets, we have witnessed the unfolding of cosmic evolution on a grand scale. We have delved into the mysteries of dark matter and dark energy, the invisible forces that shape the cosmos and drive its expansion. We have explored the lives and deaths of stars, the cataclysmic collisions of black holes, and the gravitational waves that ripple through spacetime, revealing the most violent events in the universe.

At the same time, our journey has taken us back in time to the cosmic dawn, where the first stars ignited and the universe emerged from darkness into light. We have glimpsed the quantum realm, where the laws of physics blur and spacetime itself becomes a quantum entity. We have peered into the multiverse, a vast and speculative landscape of parallel universes with different properties and laws of physics.

As we conclude our exploration, we are left with a sense of awe and wonder at the vastness and complexity of the cosmos. The mysteries that remain are as tantalizing as ever, beckoning us to continue our quest for understanding and discovery. The future of cosmology holds promise and potential, as new technologies, theoretical innovations, and interdisciplinary collaborations propel us forward into uncharted territory.

But beyond the scientific quest lies a deeper truth: that the universe is not just a collection of stars, galaxies, and cosmic phenomena, but a reflection of the human spirit—a testament to our curiosity, creativity, and capacity for wonder. In our exploration of the cosmos, we discover not only the secrets of the universe but also the beauty and majesty of the human endeavor.

As we gaze up at the night sky, let us remember that we are part of something greater than ourselves—a vast and wondrous cosmos that inspires us to reach for the stars and unlock the mysteries of the universe. The journey may be long and challenging, but with curiosity as our guide and determination as our fuel, the cosmos awaits, ready to reveal its secrets to those who dare to explore its depths.

Acknowledgments:

I would like to express my deepest gratitude to everyone who contributed to the creation of "The Universe and Cosmology: Explore the Mysteries of the Cosmos, from the Big Bang Theory to the Latest Discoveries in Astronomy and Astrophysics." This book would not have been possible without the support, encouragement, and expertise of many individuals and organizations.

First and foremost, I would like to thank the scientists, researchers, and astronomers whose pioneering work has expanded our understanding of the universe and inspired generations of explorers. Their dedication, passion, and commitment to advancing the frontiers of knowledge have laid the foundation for this book and countless other scientific endeavors.

I am also grateful to the publishers, editors, and production team who helped bring this book to life. Their professionalism, attention to detail, and tireless efforts ensured that the content was of the highest quality and met the standards of excellence that readers expect.

I would like to extend my appreciation to my colleagues and peers in the field of cosmology and astronomy for their valuable insights, feedback, and collaboration. Their expertise and guidance have enriched the content of this book and helped shape its narrative.

I am deeply thankful to my family and friends for their unwavering support, encouragement, and understanding throughout the writing process. Their love, patience, and encouragement kept me motivated and inspired, even during the most challenging times.

Last but not least, I would like to express my heartfelt thanks to the readers of this book. Your curiosity, enthusiasm, and passion for exploring the mysteries of the universe are what drive the pursuit of knowledge and discovery. It is my hope that this book will spark your imagination, ignite your curiosity, and inspire you to embark on your own cosmic journey of exploration and discovery.

Thank you to all who have contributed to this project in ways big and small. Your contributions are deeply appreciated and have made this book a reality.

www.ingramcontent.com/pod-product-compliance
Lightning Source LLC
Chambersburg PA
CBHW071206240526
45470CB00018B/1524